Basic microbiology

EDITOR: J. F. WILKINSON

Volume 4

Microbial physiology

IAN W. DAWES B.Sc., D.Phil.
IAN W. SUTHERLAND D.Sc.
Department of Microbiology, University of Edinburgh

BLACKWELL SCIENTIFIC PUBLICATIONS
OXFORD LONDON EDINBURGH MELBOURNE

© 1976 Blackwell Scientific Publications
Osney Mead, Oxford OX2 OEL
8 John Street, London WC1N 2ES
9 Forrest Road, Edinburgh EH1 2QH
P.O. Box 9, North Balwyn, Victoria, Australia

ISBN 0 632 00278 6

First published 1976

British Library Cataloguing
in Publication Data

Dawes, Ian W
    Microbial physiology. — (Basic
    microbiology; vol. 4).
    Bibl. — Index
    ISBN 0-632-00278-6
    1. Title 2. Sutherland, Ian Wishart
    3. Series
    576'.11   QR84
    Micro-organisms — Physiology

Distributed in the U.S.A. by
Halsted Press, a division
of John Wiley & Sons, Inc.,
New York

Set in Journal Roman
by Preface Ltd, Salisbury
Printed in Great Britain
by Page Bros (Norwich) Ltd
Bound by
Kemp Hall Bindery, Oxford

Contents

Preface

Most prefaces begin with 'This book is . . .', followed by an apology for its existence, a statement of the aims of the authors, an acceptance of responsibility for all errors and omissions, and a grateful acknowledgement of the patience of the authors' spouses.

The title of this book is obvious, and we have no intention of putting off potential customers by apologising for any shortcomings. Errors are inevitably the result of the other author. The size was set by our genial publisher who must therefore accept all kudos for the delightful brevity of the book, its lack of boring detail, and of course any omissions. Our wives are no more, nor less, patient than they were before we began and we remain happily married to them.

A final word of warning to our readers. This book is intended (it had to come in somewhere) as an introduction to microbial physiology to cover a broad spectrum of course requirements. It provides a foundation upon which to build by judicious reading of other sources. We trust our readers find the book useful and we would be delighted to receive any comments in case we have the opportunity to prepare another edition.

Ian W. Dawes
Ian W. Sutherland   *Edinburgh 1976*

Introduction

Physiology, according to the *Concise Oxford Dictionary*, is the science of normal functions and phenomena of living things. Even when restricted to microorganisms the present knowledge is very extensive, and like microorganisms in a nutrient culture, is increasing exponentially. There are many reasons for this interest. For some, trying to understand the basics of life processes, microorganisms have become experimental tools largely because they can be grown and manipulated with relative ease. For many, the choice of experimental organism has been just one bacterial species, *Escherichia coli*, since biochemical and genetic techniques have been developed to a level of sophistication unmatched in any other organism. There are however a number of major differences between prokaryotic and eukaryotic organisms, and a number of eukaryotic microorganisms (notably the slime mould *Dictyostelium*; the yeast *Saccharomyces*; fungi such as *Neurospora* and *Aspergillus*; algae *Chlamydomonas* and *Chlorella*, and the protozoan *Tetrahymena* are receiving increasing attention as representative eukaryotes amenable to detailed study.

    E. coli, moreover, does not typify the prokaryotic cell. Different groups of bacteria have adapted to grow in different environments to the extent that they can be isolated from almost any ecological niche, including such unexpected places as the mouths of geysers and the fuel tanks of aeroplanes. This diversity is reflected in the range of activities to be found in different groups, and in their response to particular environmental conditions. Bacteriologists, ecologists, many industrial microbiologists, taxonomists and developmental biologists are therefore more likely to be interested in the differences between various microorganisms, as well as their similarity to *E. coli* or their conformity to the notion of a 'generalised microbial cell'.

    In this book there is an attempt to strike a balance between the two approaches of studying a few organisms in depth, and discussing the different activities to be found amongst different groups of microorganisms. Where possible eukaryotes are introduced into the discussion to emphasise the differences between them and prokaryotes. It is assumed that the reader has a basic knowledge of biochemistry so that less space has been given to common biological processes, and more to those aspects in which microorganisms differ from plant and animal cells.

    The astute student will realise that any book of this size can only act as an introduction to the subject and provide a basis for further reading. For those wanting to obtain a broader coverage of the topics outlined here, or to delve into other developing areas of microbiology, a list of references to useful review articles is provided at the end of the book. In addition to these, the following review journals are very important sources of information and act as indicators of current trends in the study of microbiology: *Annual Reviews of Microbiology, Symposia of the Society for General Microbiology, Bacteriological Reviews, Advances in Microbial Physiology,* and *Critical Reviews in Microbiology.*

1 Chemical cytology of the microbial cell

Microbial cells, even those of eukaryotes, are usually so small that important structures cannot be resolved by light microscopy. Consequently, much of our present knowledge of cell structure is the result of electron microscopy. Specimens for electron microscopy often require extensive pretreatment before they can be visualised satisfactorily, and care must be taken in interpreting electron micrographs. During the preparation of thin sections many compounds are inactivated and there are problems in trying to locate some labile chemical components, such as enzymes, within particular structures by cytochemical staining.

Techniques have been developed for disrupting cells so that some organelles or other particles can be isolated in a state which retains at least some of the *in vivo* structure and activity. Particular structures can usually be isolated from cell lysates by zonal centrifugation through sucrose density gradients. Here also care should be taken in interpreting the results, since disruption of cells often involves harsh physical treatments needed to overcome the strength of the cell wall surrounding most microorganisms. Methods used to break cells include grinding with glass beads or sand, exposure to ultrasonic vibration or sudden changes in pressure, or to more gentle osmotic lysis after removing the cell wall with lytic enzymes.

Before discussing general cytological features found in microorganisms it is well to recall that there are many fundamental differences between prokaryotes and eukaryotes. These are summarised in Table 1.1, and Fig. 1.1 illustrates structures which can be found in prokaryotic and eukaryotic microorganisms. There is a very wide range of cytological features found in microbes, and those outlined in Fig. 1.1 and in the subsequent discussion need not occur in all species. There may also be extensive differences between cells of the same species grown under different physiological conditions, and where quantitative results are important they must include a definition of the culture conditions used.

CELL SURFACE AND ITS APPENDAGES

Flagella and Cilia

Cell motility is found in both eukaryotic and prokaryotic microorganisms and is usually dependent on specialised organelles protruding from the cell surface. In bacteria these are flagella; in eukaryotes the terms flagella and cilia have both been used, although the structures are essentially similar.

3

Table 1.1 Brief comparison of prokaryotic and eukaryotic cells

Characteristic	Prokaryotic cell	Eukaryotic cell
Size	Usually around $1-5$ μm	Usually greater than 5 μm
Movement	Flagellar or gliding motion, simple fibrillar arrangement of each flagellum	Flagellar or amoeboid motility, complex fibrillar arrangement of flagellum
Wall structure	Usually contains several polymers, almost always peptidoglycan	Contains a variety of organic or rarely inorganic polymers, never peptidoglycan
Vacuoles	Rarely present, if so gas vacuoles	Often present, range of different types and functions
Arrangement of nuclear DNA	No delineating nuclear membrane, single circular chromosome attached to cell membrane or mesosome	Within membrane-bound nucleus as several linear chromosomes. DNA complexed to basic proteins, histones
Replication of DNA and segregation	Bidirectional from single replication origin. Amitotic segregation	Bidirectional replication from multiple origins. Limited to part of cell cycle and segregation by mitosis or meiosis
Protein synthesis	Translation simultaneous with transcription. Only one RNA polymerase known, with modifying proteins. Ribosomes are 70 S and inhibited by a group of antibiotics specific for prokaryotes.	Translation of nuclear genes occurs in cytoplasm. Three RNA polymerases. Cytoplasmic ribosomes are 80 S and inhibited by cycloheximide. Mitochondrial and chloroplast ribosomes resemble those of prokaryotes
Energy production	Respiration, fermentation or photosynthesis. Wide range of substrates. Respiratory chain associated with plasma membrane, Photosynthesis on invaginations.	Some fermentative, but usually either respiratory or by photosynthesis. Respiration in mitochondria, photophosphorylation in chloroplasts
Reproduction	Asexual, by binary fission. Conjugation is rare, leads only to partial diploids and is not associated with reproduction	Sexual or asexual. Many ways, including budding, binary fission, hyphal extension and sporulation. Conjugation part of reproduction and leads to diploids

Bacterial Flagella

Flagella are helical structures several times the length of the bacterial cell (Fig. 1.2) and if sheared off by ultrasonic treatment motility is lost, but is regained as new flagella are resynthesised. The pitch and wavelength, and the arrangement of flagella on the cell surface are characteristic to each species. In *Pseudomonas* species there may be one or two polar flagella, while in other genera they may be found in large numbers over the whole surface of the cell. In *spirochaetes* a modified type of flagellum is present as an *axial filament*; two sets of fibrils are fixed at the poles of

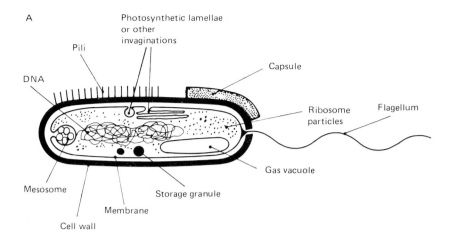

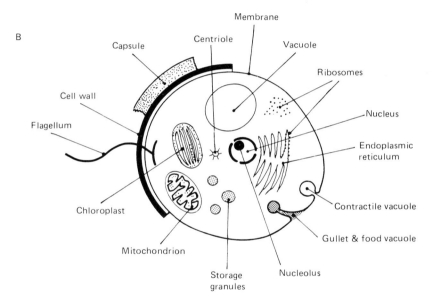

Figure 1.1 'Representative' A, prokaryotic; and B, eykaryotic cells. Some structures shown are dispensable, including capsules, flagella, membrane invaginations, pili, storage granules, vacuoles and chloroplasts. Obviously these do not represent form and shape of cells, merely the main structures.

the cell and run the length of the cell enclosed within the outermost layer of the cell surface. This arrangement enables these spirally shaped organisms to move by a flexing motion.

Flagella are composed of subunits of a single protein, *flagellin*, which in the electron microscope are seen to be aggregated into chains of molecules arranged helically. Flagellins from different species differ in their amino acid composition

5

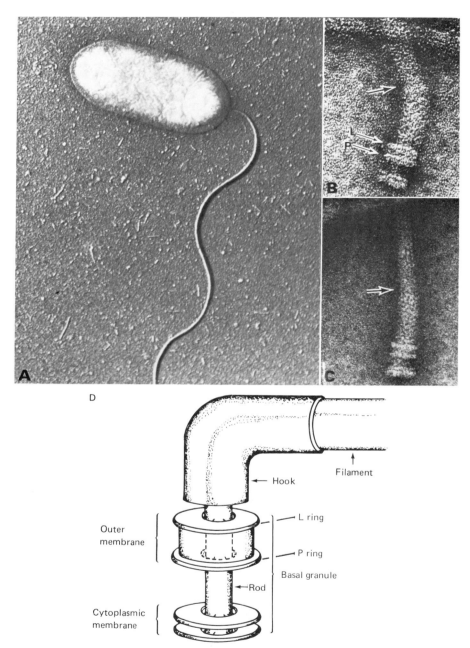

Labels in figure D: Hook, Filament, L ring, Outer membrane, P ring, Basal granule, Rod, Cytoplasmic membrane

Figure 1.2 The bacterial flagellum. A, Cell with a single polar flagellum, electron micrograph of a shadowed cell. B, Basal granule, negatively stained. C, Some subunit structure of the flagellum revealed by negative staining. D, Interpretation of the basal body structure. (A, by courtesy of Professor J. P. Duguid; B, C and D, by courtesy of Dr. Julius Adler and the *Journal of Bacteriology*)

and serological properties and in some species (*Salmonella* species and *Spirillum serpens*) contain an unusual amino acid, ε-*N*-methyllysine.

Flagellar assembly provides a simple and interesting model of how biological structures are assembled from their chemical components. Under acid conditions *Bacillus pumilis* flagellar filaments can be dissociated into their subunit polypeptide of molecular weight 30 000 to 40 000. On slowly raising the pH these reaggregate to give straight filaments as well as the wavy flagellum-like structure. The straight form undergoes rearrangement to the more stable wavy form. Experiments using *p*-fluorophenylalanine, which leads to formation of abnormal flagella, have shown that they grow by condensation of subunits at the tip distal to the cell membrane, and raises the intriguing question of how the subunits are transported there — through the hollow central core of the flagellum?

The filament of each flagellum is attached to the cell membrane at a structure known as the *basal granule* or disc. The precise structure varies between species, but a typical example is shown in Fig. 1.2. A filament is joined by a hook to a rod and set of rings located in the cell wall and the cell membrane. This basal granule is probably involved in the transduction of energy from the cytoplasm or membrane to the flagellum. The complete assembly and functioning of the bacterial flagellum is quite complicated; in *E. coli* and *Salmonella typhimurium* some twenty genes are involved including that for flagellin, and those determining rotation of the 'motor' including whether or not it turns the flagellum, and if so, whether it can go both clockwise and anticlockwise.

Eukaryotic Flagella and Cilia

Eukaryotic flagella or cilia are much larger and more complex than the corresponding prokaryote organs and are always attached to cylindrical basal bodies within the cytoplasm. An extension of the plasma membrane surrounds the flagella and encloses a system of microtubules (each resembling a bacterial flagellum) arranged as two central ones surrounded by a further nine pairs (Fig. 1.3). The only

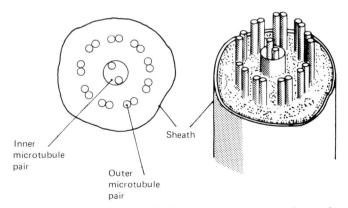

Inner
microtubule
pair

Sheath

Outer
microtubule
pair

Figure 1.3   Crosssection   and   diagrammatic   representations   of   a eukaryotic flagellum

7

difference between flagella and cilia lies in their length, up to 200 μm and up to 10 μm respectively, and in the much greater number of cilia found on the cell.

Mechanism of Motility

How do flagella move cells? In electron micrographs they appear as sinusoidal structures, but this is really a two-dimensional visualisation of a three-dimensional shape. There are several ways such helices can move cells, and it appears that eukaryotes and bacteria differ in this respect. Eukaryotic flagella contain complicated bending machinery, and an input of energy at one end, causing a slight perturbation of flagellin subunits at the base, leads to a helical wave travelling the length of the flagellum.

Bacterial flagella, on the other hand, appear to rotate rigidly. When the distal end of a straight filament is fixed to a glass slide with an anti-filament antibody, the cell rotates at several revolutions per second. Similarly, after fixing polystyrene beads to mutant cells with a straight filament the beads rotate about the axis of the filament in one direction while the cells rotate in the other (Fig. 1.4).

At the moment it is not clear how energy is supplied to either system, although in eukaryotes ATP may be involved since detached eukaryotic flagella beat if ATP is added. This ATP is hydrolysed by an ATPase thereby affording a source of energy. Bacterial flagella do not possess ATPase activity, and although ATP does cause an interconversion between two arrangements of flagellin subunits isolated from *S. typhimurium*, it is not hydrolysed. The bacterial 'motor' may well reside in the cell membrane with the basal body acting as an energy transducer. Located in the bacterial membrane is the respiratory chain generating energy for transport across membranes and oxidative phosphorylation of ADP to ATP (pp. 64, 83) and

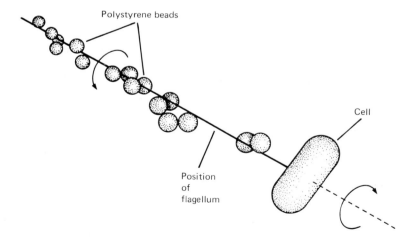

Polystyrene beads

Cell

Position
of
flagellum

Figure 1.4 Rigid rotation of the bacterial flagellum shown by fixing polystyrene beads to the straight flagellum of an *Escherichia coli* mutant. (By courtesy of Dr. H. C. Berg and *Nature*)

this also generates the energy for flagellar rotation. This coupling of energy released in the respiratory chain does not require ATP as an intermediate; rather, as we will see later, solute and ion transport, ATP biosynthesis and flagellar rotation are alternatives for linking into the respiratory chain.

Most bacteria move steadily in almost a straight line, then alter course abruptly. In some organisms, including *E. coli* and photosynthetic bacteria, the choice of a new direction seems random, but can be influenced by chemicals (or for photosynthesisers, light) acting as attractants or repellants.

Another form of motility, gliding movement, is found in blue-green algae and related bacteria, and in myxobacteria. The cells move slowly over solid surfaces although no recognisable organs of locomotion can be seen. The only common property found in such cells which might be involved in this movement is the secretion of mucilage over the exterior of the cells and onto the surface of the medium. In eukaryotes *amoeboid* movement is found in slime moulds and some protozoans. This is the result of *cytoplasmic streaming* in organisms without a rigid cell wall, and occurs on solid surfaces.

Pili and Fimbriae

At the surface of many Gram-negative bacteria may be found numerous filamentous appendages called *pili* or *fimbriae* which are up to 3 μm long and have a diameter of 5 to 10 nm (Fig. 1.5). A second type of pilus is found in several bacteria which undergo conjugation; this 'sex pilus' is much larger than common 'type I' pili or fimbriae. It is 25 to 30 nm thick and intermediate in length between common fimbriae and flagella. Only one or two sex pili are present on each cell.

Pili or fimbriae resemble flagella in being composed of a protein which can be disaggregated and show spontaneous self-assembly, but they are readily distinguishable from flagella by their smaller diameter and absence of wave structure. All types of pili are hollow; during bacterial conjugation genetic material passes through a sex pilus from donor to recipient cell. The sex pili are also the site of adsorption of a group of highly specialised bacteriophages, some of which adsorb at the tip and others along the length of the pilus. Pili or fimbriae may play a part in adhesion of bacterial cells either to other cells or particulate material, which is obviously important for the utilisation of solid substrates and in the much more specialised process of bacterial conjugation and transfer of genetic material. They also provide a means of attachment in aqueous environments.

Capsules and Slime

Many microorganisms have a discrete external layer of mucilaginous material called a *capsule* which completely surrounds the cells and occludes the cell walls. The size of the capsules and the amount of capsular material produced is markedly dependent on the cultural conditions and is often favoured by a high degree of aeration and a high carbon to nitrogen ratio in the growth medium. In a few

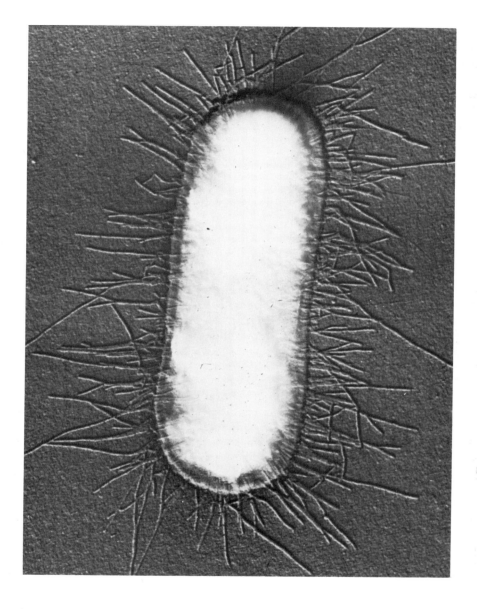

Figure 1.5 Pili at the surface of *Escherichia coli*. Shadowed electron micrograph. (By courtesy of Professor J. P. Duguid.)

Table 1.2   Chemical composition of microbial capsules and slime

Organism	Nature of polymer	Capsule/slime	Monomers
Bacillus anthracis	Polypeptide	Capsule	D-glutamic acid
Leuconostoc mesenterioides	Homopolysaccharide (dextran)	Slime	D-glucose
Streptococcus pneumoniae type 3	Heteropolysaccharide	Capsule or slime	D-glucose, D-glucuronic acid
Klebsiella aerogenes type 54	Heteropolysaccharide	Capsule or slime	D-glucose, L-fucose, D-glucuronic acid
Escherichia coli strain K12	Heteropolysaccharide	Slime	D-glucose, D-galactose, L-fucose, D-glucuronic acid, acetate, pyruvate
Cryptococcus neoformans	Heteropolysaccharide	Capsule	D-xylose, D-mannose, D-galactose, D-glucuronic acid

species, there is no capsule and only cell-free slime is formed; capsules and slime in most microbial species are polysaccharides, termed *extracellular* polysaccharides to distinguish them from others found within the cell walls.

Microbial extracellular polysaccharides may be homopolysaccharides formed from a single sugar, or heteropolysaccharides composed of two or more sugars. A large number of different monosaccharides have been identified in capsular and slime polysaccharides including: neutral sugars — D-glucose, D-galactose and D-mannose (all hexoses); L-fucose and L-rhamnose (methylpentoses); amino sugars — N-acetyl-D-glucosamine and N-acetyl-D-galactosamine; and uronic acids — D-glucuronic acid and D-galacturonic acid (Table 1.2). Pentoses such as D-ribose and D-xylose are seldom found in bacterial capsules and slime but are frequently components of the extracellular polysaccharides of yeasts and algae. In addition, numerous extracellular polysaccharides contain combined phosphate, acetate, formate or pyruvate.

There may be many possible functions of microbial capsules but non-capsulate variants grow as well if not better under laboratory conditions as capsulate wild-type strains. In its natural environment a capsule may, for instance, protect an organism against desiccation, phage infection or phagocytosis. For pathogenic bacteria, capsule formation is associated with virulence, affording protection against attack by both antibodies and by macrophages.

Cell Walls

The shape of most microbial cells is due to the presence of a rigid cell wall, which has the major function of protecting the fragile protoplast (cell membrane and its contents) from osmotic lysis. Microorganisms usually live in an environment which is hypotonic to the cell cytoplasm, and unless the protoplast is supported by the cell wall it expands and lyses. The rigidity of the cell wall is mainly conferred by a single polysaccharide or related component; other polysaccharides are present and these are generally characteristic to particular groups of organisms or even strains of one species. The polymeric wall components are organised into complicated multi-layer structures; some bacterial examples are shown in Fig. 1.6.

The wall has other functions, the highly charged polymers may provide an ion-exchange mechanism assisting in the uptake of ions and nutrients. It is also effectively a molecular sieve providing a barrier to entry of some molecules, and retaining proteins found in the *periplasm*, the region between the wall and membrane in Gram-negative bacterial cells.

Prokaryotic Cell Wall

Apart from *Mycoplasma* species, which lack cell walls, and *Halobacteria*, which have an unusual wall structure, bacteria and blue-green algae can be grouped into one of two groups, Gram-positive or Gram-negative, according to their wall structure and composition. This difference was initially based on the ability of the organisms to retain crystal violet during Gram's staining procedure, but with electron microscopic and chemical analysis of wall structures it became clear that the staining reflected an intrinsic difference in the chemistry and arrangement of the wall layers between the two groups.

Cytological differences between Gram-positive bacteria and Gram-negative bacteria are indicated in Fig. 1.6. Gram-positive walls are thicker and appear relatively amorphous, whereas Gram-negative 'envelopes' are much more complicated and show a multi-layered structure. This is seen also in the chemical composition of the walls (Table 1.3) and in the localisation of polymers (Fig. 1.6). Several of the polymers found in bacterial cell walls are unique to prokaryotes. These are discussed below.

Peptidoglycan   Peptidoglycan is found in almost all bacteria (except *Mycoplasma* and *Halobacteria*) and is not only unique in itself to the prokaryotic cell, but contains up to three components not found in eukaryotic polymers: D-amino acids, muramic acid and diaminopimelic acid. Peptidoglycan is the polymer conferring rigidity on the bacterial cell, determining shape and resistance to osmotic lysis. When peptidoglycan is removed by lysozyme in hypotonic medium the cells lyse. In hypertonic media, rounded spheroplasts or protoplasts are formed depending on whether part or all of the wall is removed. Peptidoglycan is essentially a linear polymer of alternating residues of N-acetyl-D-glucosamine and N-acetyl-D-muramic acid with separate chains crosslinked to varying degrees by short peptide bridges. The polysaccharide backbone of the molecule shown in Fig. 1.7 is ubiquitous to prokaryotes containing the polymer. Attached to the lactyl moiety of the muramic acid residues are peptide side chains containing four amino acids in a characteristic sequence, L-alanyl-D-glutamyl-X-D-alanine; X may be neutral (e.g. homoserine) but

12

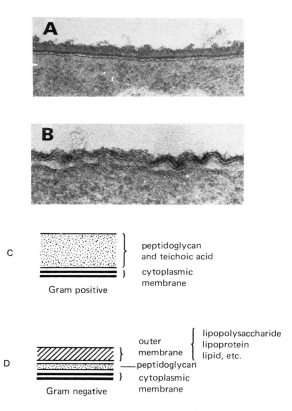

Figure 1.6 Cell wall structures seen in thin-section electron microscopy. A, Gram-positive *Bacillus subtilis* (x 60 000); B, Gram-negative *Escherichia coli* (x 120 000); C, Diagrammatic representation of the Gram-positive wall; D, of the Gram-negative wall. The location of wall components is indicated. (Electron micrographs by courtesy of Dr. P. J. Highton).

is more commonly a diamino acid such as L-ornithine, L-lysine or diaminopimelic acid as the LL meso-isomer.

$$\underset{\text{HOOC}}{\overset{\text{NH}_2 \ (\text{L})}{\text{CH}}}-\text{CH}_2-\text{CH}_2-\text{CH}_2-\underset{\text{NH}_2}{\overset{(\text{D or L})\ \text{COOH}}{\text{CH}}}$$

diaminopimelic acid

In Gram-negative bacteria, the peptidoglycan is probably present as a single layer and the chains are not crosslinked to a very great extent, see Fig. 1.6. In Gram-positives, the wall appears to be composed of multiple layers of peptidoglycan chains (about twenty in *Bacillus subtilis*), and, particularly in cocci,

13

Table 1.3 Composition of microbial cell walls

BACTERIA

Organism type	Peptidoglycan	Teichoic acid	Lipopolysaccharide	Lipid	Protein
Gram-positive	40–50%	+	–	2%	c. 10%
Gram-negative	5–15%	–	+	20%	c. 60%

FUNGI

Organism group	Wall components
Acrasiales	cellulose
Oömycetes	cellulose, glucan
Zygomycetes	chitosan, chitin
Ascomycetes (most)	chitin–glucan
Basidiomycetes (most)	chitin–glucan
Deuteromycetes	chitin–glucan
Saccharomycetaceae	mannan, glucan and chitin in bud scars
Sporobolomycetaceae	mannan, chitin

the backbone chains are extensively crosslinked. Crosslinkage occurs between the carboxyl group of the terminal D-alanine residue and either a free amino acid of the tetrapeptide, or the terminal amino acid in an additional peptide sequence. The bridge between the two chains can thus be: (i) a direct peptide bond (*E. coli*); (ii) by incorporation of a single additional amino acid (many Gram-positives); (iii) by a peptide of up to five amino acid residues, commonly of glycine but sometimes also containing serine or alanine (many Gram-positives); or, (iv) by an extra peptide with essentially the same composition as the tetrapeptide already attached to a muramic acid residue (*Micrococcus lysodeikticus*). The whole cell wall peptidoglycan in some bacteria may be envisaged as a giant net-like molecule of considerable strength.

Teichoic acids These unusual anionic polymers of glycerol phosphate, ribitol phosphate (Fig. 1.8) or other sugar phosphates, are found in Gram-positive cell walls and membranes. Their precise function is not clear, although they may provide an essential negatively charged environment in the cell wall for binding of ions such as Mg^{2+} or for regulating enzyme activity. When phosphate is a limiting component in the medium, teichoic acids are replaced by other negatively charged polymers, the teichuronic acids which are composed of glucuronic acid and N-acetylgalactosamine. Teichoic acids bind to autolytic enzymes present in the cell walls, and may also be involved in regulating the lytic modification of the cell wall peptidoglycan necessary for normal cell division (see Chapter 2).

Lipopolysaccharides Lipopolysaccharides form one component of the outer membrane of Gram-negative bacteria. They are very variable in chemical structure

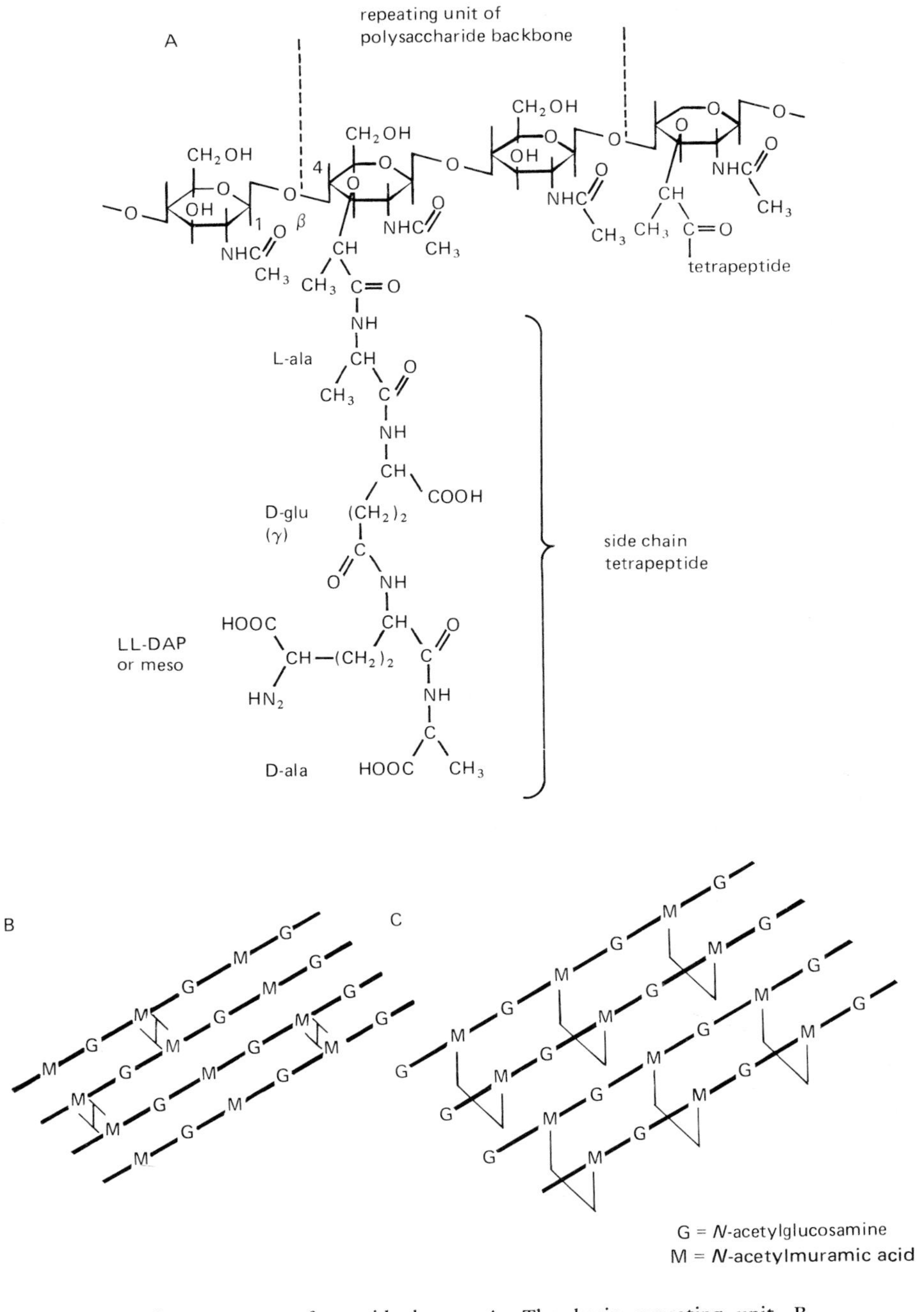

Figure 1.7 The structure of peptidoglycan. A, The basic repeating unit, B, crosslinking in Gram-negative *Escherichia coli* and C, Gram-positive *Staphylococcus aureus*

15

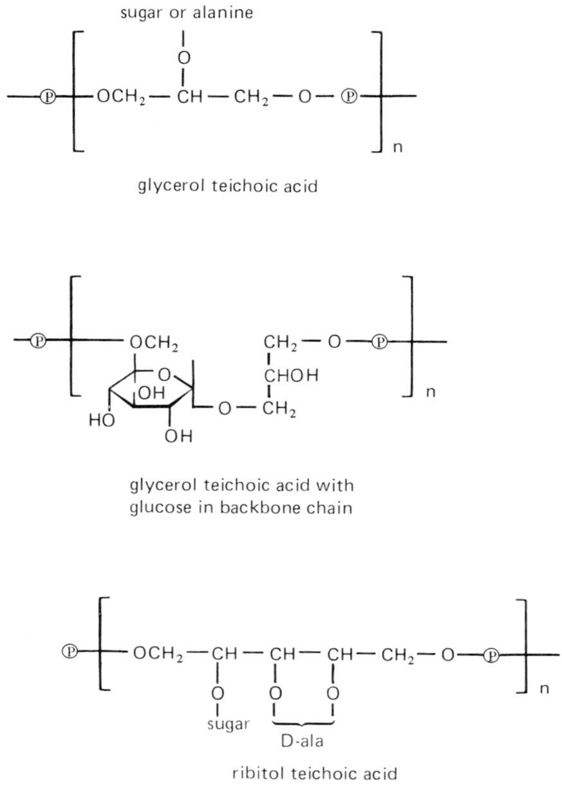

sugar or alanine

glycerol teichoic acid

glycerol teichoic acid with
glucose in backbone chain

ribitol teichoic acid

Figure 1.8   Teichoic acids

from one group of organisms to another, but usually contain between five and nine different sugars, including some not found in other microbial polymers (e.g. 3,6-dideoxyhexoses) and a lipid (Lipid A) which has the unusual structure shown in Fig. 1.9. Almost all enterobacterial lipopolysaccharides contain D-glucose, D-galactose and N-acetyl-D-glucosamine, as well as two sugars which are unique to lipopolysaccharides — L-glycero-D-mannoheptose and 2-keto-3-deoxyoctonate

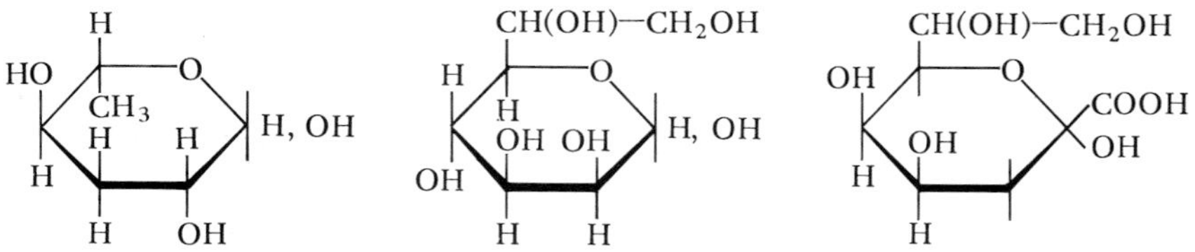

3,6-dideoxyhexose   L-glycero-D-mannoheptose   2-keto-3-deoxyoctonate

The function of the lipopolysaccharide is not known, and mutants lacking the O-antigens and much of the core are viable under laboratory conditions. The

16

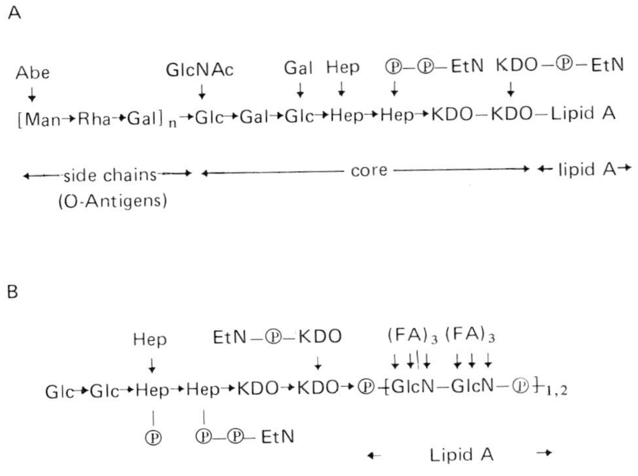

A

Abe   GlcNAc   Gal Hep P–P–EtN KDO–P–EtN
↓   ↓   ↓   ↓   ↓   ↓
$[Man{\rightarrow}Rha{\rightarrow}Gal]_n{\rightarrow}Glc{\rightarrow}Gal{\rightarrow}Glc{\rightarrow}Hep{\rightarrow}Hep{\rightarrow}KDO{-}KDO{-}Lipid\ A$

←——side chains——→ ←——————————— core ———————————→ ← lipid A→
(O-Antigens)

B

Hep   EtN–P–KDO $(FA)_3\ (FA)_3$
↓   ↓   ↓↓|↓ ↓↓↓
$Glc{\rightarrow}Glc{\rightarrow}Hep{\rightarrow}Hep{\rightarrow}KDO{\rightarrow}KDO{\rightarrow}P{+}GlcN{-}GlcN{-}P\}_{1,2}$
| |
P P–P–EtN
                 ←   Lipid A   →

Man: mannose; Glc: glucose; GlcNAc: *N*-acetylglucosamine;
Rha: rhamnose; Gal: galactose; Hep: heptose; KDO: ketodeoxyoctonate;
Abe: abequose; GlcN: glucosamine; EtN: ethanolamine;
P: phosphate; FA: fatty acids (usually β-hydroxymyristic acid).

Figure 1.9 Lipopolysaccharides of A, *Salmonella typhimurium* and B, *Escherichia coli* B

complete structure is, however, necessary for pathogenicity, and different parts of it form recognition sites for absorption of different bacteriophages and bacteriocins.

Eukaryotic Cell Wall

Microbial eukaryotes are more diverse in their cell wall structures, ranging from the silica of siliceous diatoms to a range of polysaccharides in fungi. The walls of fungi have been studied in some detail: the entire spectrum of different fungi can be separated into categories on the basis of the chemical nature of their cell walls. These categories follow conventional taxonomic boundaries closely. Fungal cell walls are composed of polysaccharides (80—90%) including cellulose, glucans, chitin, mannans and galactans. The vast majority of fungi, including all forms with typical septate mycelium have a chitin-glucan wall, whereas yeasts have an increased mannan component. In fungi exhibiting dimorphism (the ability to change from a mycelial to a yeast-like habit), this change is accompanied by a change in the relative amounts of mannan and chitin in the cell walls. The shape of a cell is therefore related to the type of polymer present in wall — although there is undoubtedly a contribution from the spatial arrangements of sites for wall synthesis

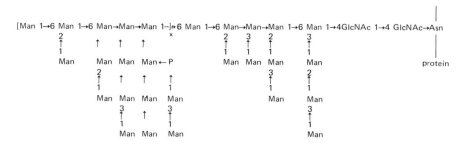

Figure 1.10 Suggested structure for the mannan of a strain of *Saccharomyces cerevisiae* and *Kluyveromyces* species

in the cell membranes. In filamentous fungi the cell wall grows by deposition mainly at the hyphal tip, whereas in yeasts, new wall material is inserted more diffusely with a concentration of synthesis near the base of the bud. Another indication that cell wall components play an important part in structural differentiation is the fairly specific association of chitin with bud scars in yeast.

Electron microscope studies show that the fungal wall is a fabric of interwoven microfibrils embedded in, or cemented by, amorphous components. The microfibrils are usually chitin, a $\beta(1 \to 4)$ linked polymer of N-acetylglucosamine, or cellulose, or glucan which is a highly branched polymer of glucosyl residues linked $\beta(1 \to 6)$ and $\beta(1 \to 3)$. This arrangement is seen by partially digesting the cell wall. The amorphous components of yeasts are mannans; these are the immunodeterminant constituents of the cell wall. Mannans of two yeasts are indicated in Fig. 1.10. In some cases the mannans are phosphorylated and linked to proteins.

THE PROTOPLAST

Membranes

The cytoplasmic membrane of bacteria performs several functions which in eukaryotes are located in organelles. For this reason we will consider the two groups separately. Before doing so, however, the general concepts of membrane structure should be considered. These probably apply to most, but not all, membranes found in living systems.

General Membrane Structure

In thin sections examined by electron microscopy the membrane has a 'unit membrane' structure: two densely staining layers separated by a non-staining region (Figs. 1.6 and 1.13). The classical concept of Davson and Danielli suggests that this unit structure is a bilayer of phospholipid molecules with polar groups facing outwards into aqueous phases at the membrane surfaces, and the fatty acyl moieties

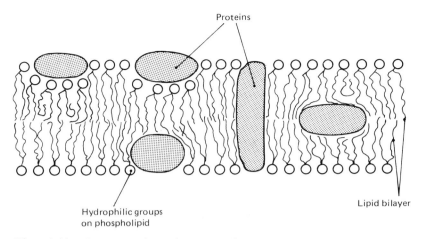

Figure 1.11 Structure of membranes, a diagrammatic representation. The lipid molecules are probably in constant motion

forming a hydrophobic, semi-liquid phase to the interior (it should be noted that membranes contain about 40% phospholipid). A variety of physical studies of *Mycoplasma* membranes in particular tends to support this view of the membrane as a permeability barrier to ions and polar molecules.

Proteins also play an important part in membrane structure. Much of the protein is apparently associated with the surfaces, but freeze-etch electron microscopy indicates that there may be particular regions of the membranes which have protein molecules embedded in, or even across, the lipid bilayer.

The membrane is regarded as a fluid mosaic of small functional regions that are predominantly protein (approximately 60% of the membrane by weight) embedded in a relatively homogeneous phospholipid bilayer matrix (Fig. 1.11). Hydrophobic bonds are thought to predominate and the phospholipid polar regions may largely be exposed. Many different proteins sediment with membranes during centrifugation. Some are enzymes bound to a greater or lesser extent to lipid. When proteins of intact cell membranes are chemically labelled by reaction with a molecule to which membranes are impermeable some polypeptides are labelled, others are not unless the membrane has been ruptured. This indicates that membranes have a distinct polarity with different proteins accessible to the aqueous environment on opposite sides. In the next chapter we will see that there is functional as well as structural polarity shown by the vectorial transport of compounds into and out of the cell.

The above concept of membrane structure has been confirmed in an elegant, and technically very sophisticated electron microscopic study of the purple membrane of the halophilic bacterium, *Halobacterium halobium.* This specialised membrane functions as a light-driven proton pump and is unusual in containing identical (rhodopsin) protein molecules of 26 000 daltons molecular weight comprising 75% of the total mass. These proteins have been visualised in a three-dimensional map of

19

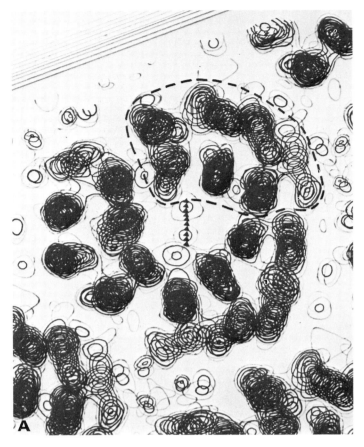

Figure 1.12A The purple membrane of *Halobacterium halobium*. Three-dimensional potential map of a region 5 nm thick spanning the membrane, showing protein molecules grouped around a three-fold axis. One of these is indicated by the broken line. (Courtesy of Drs. R. Henderson and P. N. T. Unwin and *Nature*)

the membrane at 0.7 nm resolution, revealing the location of protein and lipid components, as shown in Fig. 1.12. The protein molecules are arranged in groups of three, each molecule extends *across* the membrane contacting the aqueous solvent on both sides, and the space between protein molecules appears to be packed with phospholipid to form the classical bilayer configuration.

Prokaryotic Membrane

Chemically the lipid component differs from that of eukaryotes in containing branched chain and cyclopropane-containing fatty acids. Lecithin is rarely found in the phosphatides and sterols are normally absent. In Gram-positive bacteria it is easy to isolate membranes since the outer wall can be removed with lysozyme or

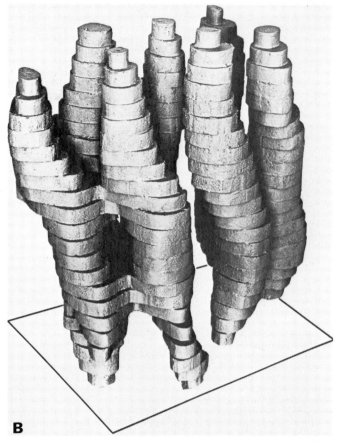

Figure 1.12B The purple membrane of *Halobacterium halobium.*
Model of a single protein molecule in the membrane, the top and
bottom are exposed to the solvent, the rest is in contact with lipid.
(Courtesy of Drs. R. Henderson and P. N. T. Unwin and *Nature*)

other lytic enzymes to leave the osmotically fragile protoplast. Gram-negative
bacteria are more resistant to lysozyme, although fragile *spheroplasts* can be
obtained by treatment with lysozyme in the presence of chelating agents. The
membrane and wall together form a complex structure which can with difficulty be
separated into component layers, including cytoplasmic and outer membrane
fractions.

Membrane Invaginations   Although the bacterial cytoplasm is completely
surrounded by the cell membrane, and lacks intracellular membrane-bound
organelles, electron microscopy may reveal numerous invaginations. These can be at
the surface or relatively deep in the cytoplasm. Those most commonly seen are
called *mesosomes,* and are found in both Gram-positive and Gram-negative cells.

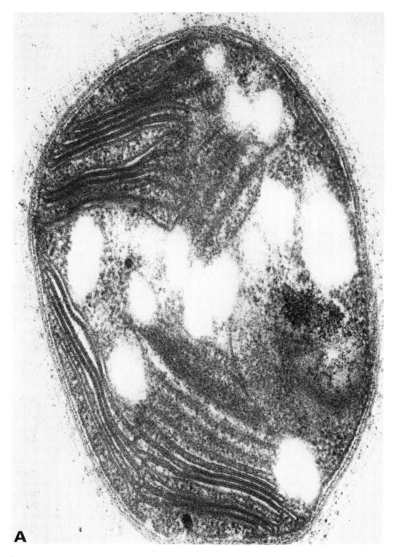

A

Figure 1.13A   Membrane invaginations in methane-utilising bacteria. Thin section. (Courtesy of Dr. S. Watson and Professor J. F. Wilkinson)

Their occurrence in the latter is dependent on the growth conditions. Mesosomes are located usually in association with the bacterial DNA or with sites of septum formation, and may therefore be involved in the organisation of chromosome segregation and cell division (see p. 36). When protoplasts of *Bacillus subtilis* are formed under slightly hypotonic conditions mesosomes are evaginated and the nuclear material is dragged to the cell membrane. Mesosomes have variable

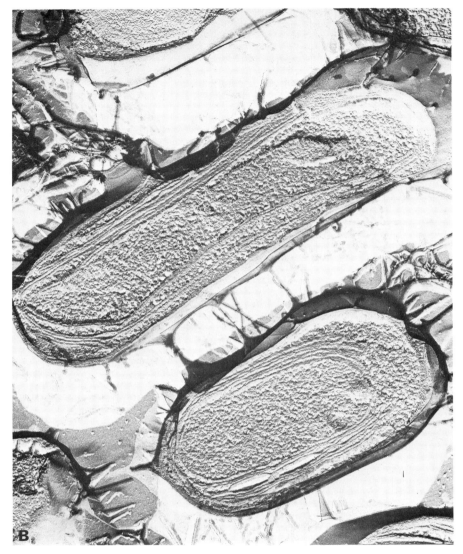

Figure 1.13B Membrane invaginations in methane-utilising bacteria. Freeze-etch electron micrograph. (Courtesy of Dr. S. Watson and Professor J. F. Wilkinson)

structures in electron micrographs, although this is probably due to changes induced in a basic structure by the fixation and embedding methods used. Despite their contiguity with the cell membrane they differ from it in their lipid and protein composition, including their enzyme and cytochrome complement.

Much more complex membrane invagination systems are found in autotrophs and specialised groups of bacteria such as nitrifying bacteria (e.g. *Nitrocystus* species) and photosynthetic bacteria (see p. 87) as well as in the nitrogen-fixing genus *Azotobacter*, methane-utilising bacteria (Fig. 1.13) and some other species. Membrane invaginations in these bacteria provide a greatly increased effective surface area to the cell which may be of particular importance to those species using gaseous substrates poorly soluble in water (e.g. oxygen, nitrogen and methane). They may also facilitate the maintenance of reducing or anaerobic conditions at the site of these reactions.

In photosynthetic bacteria, internal membrane structures called thylakoids, are arranged either as sac-like vesicles (green sulphur bacteria) or as tubular or lamellate invaginations of the cell membrane (purple sulphur, purple non-sulphur bacteria and blue-green algae). These membranes contain the complete photosynthetic apparatus (chlorophylls, carotenoids, cytochromes etc.) and their extent within the cell varies inversely as the intensity of light falling on the cells.

Functions of the Bacterial Membrane   The bacterial membrane has many important functions; some are essential to the maintenance of cell viability, while others are involved in the processes of growth and division of the organism:

Maintenance of osmotic gradients and transport of solutes   The cell membrane is not simply a semi-permeable membrane preventing the passage of high molecular weight compounds. It is selectively permeable, and is the site at which some low molecular weight substrates are transported into or out of the cell, often against a concentration gradient. Other compounds are almost totally excluded. The transport can be so specific that two structurally related sugars have very different entry rates into the cell. Moreover, the membrane has a definite polarity; while some molecules are taken up, the exit of others is facilitated. This is discussed in more detail in Chapter 3.

Organisation of cell wall synthesis   Enzymes involved in synthesising precursors of cell wall polysaccharides are associated with the membrane. These precursors are transported outward across the membrane covalently bound to an isoprenoid component of the membrane lipid. Since cell wall synthesis is localised in regions of the cell surface (see p. 36) it is possible that this reflects organisation by the membrane.

Attachment and segregation of DNA, cell division   The possible role of mesosomes in these processes has been inferred. Other aspects of this are discussed on p. 37.

Site of oxidative metabolism and energy generation   The biochemical processes associated with ATP biosynthesis in non-photosynthetic bacteria (oxidative phosphorylation and the respiratory chain, see pp. 81—84) and also ion and solute transport, are organised as multi-enzyme-protein-cofactor complexes. Stalked particles or granules (similar to those found on the inner mitochondrial membrane) containing the ATPase activity associated with oxidative phosphorylation are scattered randomly over the entire inner surface of the bacterial membrane.

Flagellar attachment   This has already been discussed (see pp. 4—8).

Eukaryotic Plasma Membrane

The plasma membranes of eukaryotes usually do not perform as many functions as those of the bacterial cell. They are thus not directly concerned with oxidative phosphorylation or organisation of DNA, these activities being associated with membranes of organelles, but they function in osmotic regulation, nutrient uptake and in some cases wall biosynthesis. In fungi, growth is restricted to the hyphal tip, and it is not clear to what extent the cytoplasmic membrane is involved. In the cytoplasm below the hyphal tip a large number of membrane-bound vesicles are found. These may contain enzymes and wall precursors which are synthesised within the cytoplasm and are released by fusion of the vesicle with the cell membrane. In eukaryotes the export of macromolecules, including extracellular enzymes, probably occurs by synthesis at the endoplasmic reticulum into vesicles and release by fusion with the external membrane.

Endoplasmic reticulum   This is a vast network of membranes within the eukaryotic cell, and is probably continuous with both the cell and nuclear membranes. It is in part covered by closely associated ribosomes and one of its functions may be to organise sites of protein sythesis. It is not known why some protein synthesis occurs on the endoplasmic reticulum, nor how much, but it has been suggested that it may be involved in the transport of messenger RNA from the nucleus. It is also associated with a number of enzyme activities not directly related to protein synthesis.

Nucleus

Thin section electron micrographs of bacterial cells reveal a region of poorly staining fibrous material. This is sensitive to DNase and is composed of DNA fibres forming what is often called a *nucleoid*. Unlike eukaryotic cells there is no nuclear membrane. By very gentle lysis of [^{3}H]-thymidine-labelled cells of *E. coli* the DNA has been shown by autoradiography to occur as a single circular (or branched circular) molecule of about 1100 μm (compare the length of the *E. coli* cell at 1 to 2 μm). The branched circular structures are replicating forms (see p. 117). This physical arrangement was predicted from genetic studies which indicated the *E. coli* chromosome was a single circular molecule. So far the few prokaryotic organisms which have been studied in sufficient detail genetically (including Gram-positive *Bacillus subtilis*, Gram-negative *Salmonella typhimurium* and filamentous *Streptomyces coelicor*) conform to this general pattern of having one circular linkage group.

In bacteria there are apparently none of the very basic proteins called histones which are associated with the DNA of eukaryotes. Instead there may be two basic amines, *spermine* and *spermidine*; as yet the function of these molecules is not known.

25

In some bacteria, extrachromosomal DNA is present as plasmids and correlates with cytoplasmic genetic elements. These include sex factors (F) and colicinogenic and drug resistance transfer factors (R). These are usually circular DNA molecules much smaller than the bacterial chromosome. They can be separated from nuclear DNA either by virtue of differences in buoyant density or by their fairly specific interaction with dyes capable of intercalating between bases in DNA (acridines or ethidium bromide).

The nuclear DNA of eukaryotic microorganisms is enclosed within a nuclear membrane (except during meiosis and mitosis). In a number of microorganisms, including yeasts, meiosis and mitosis occur atypically with no dissolution of this membrane. In some protozoa, the ciliates, (e.g. *Tetrahymena* and *Paramecium*) there are two different types of nuclei in each cell. One, the *macronucleus*, appears to divide amitotically and is concerned with gene expression. The *micronucleus* is inactive during vegetative growth, but is involved in the sexual process of meiosis. It divides by mitosis during cell division.

The nucleus is surrounded by a double unit membrane structure, each layer with some degree of selective permeability in isolated nuclei. The nuclear membranes are more permeable than those of the cell, possibly due to the presence of relatively large pores (Fig. 1.14).

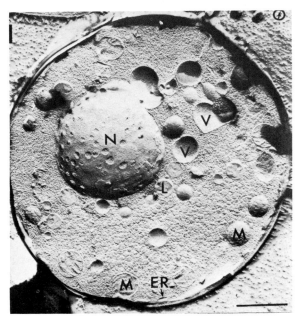

Figure 1.14 Freeze-etch electron micrograph of a *Saccharomyces cerevisiae* cell. Nuclear pores are prominent. N; nucleus; V; vacuole; M; mitochondria; L; lipid granule; ER: endoplasmic reticulum. (Courtesy of Dr. S. F. Conti and the *Journal of Bacteriology*)

The DNA of eukaryotes is arranged into a number of chromosomes, which can vary from four per haploid cell of the yeast *Hansenula* through seven in *Neurospora* to seventeen in *Saccharomyces*. These chromosomes are seen to be linear from both cytological and genetic studies. At one point on each chromosome there is a *centromere* which is the point of attachment of the chromosomes to the *spindle apparatus* formed during mitosis and meiosis for segregation of the replicated chromosomes into daughter nuclei. The eukaryotic chromosome usually appears heterogeneous, having regions of highly staining heterochromatin (which is apparently genetically inert) and less densely staining euchromatin. Histones are a group of arginine- or lysine-rich proteins which bind firmly to DNA. So far their function is not known; they may be involved in the structural organisation of the chromosome, or even in the control of gene expression. This latter function may however be carried out by other non-histone proteins present in the nucleus.

The nucleus is commonly seen to contain one or more densely staining areas, *nucleoli*. By a combination of autoradiographic, DNA-RNA hybridisation and direct electron microscopic studies the nucleolus can be shown to be the site of synthesis of ribosomal RNA (rRNA) on highly repetitive sequences of rDNA. In larger eukaryotic microorganisms the nucleolus is associated with a particular

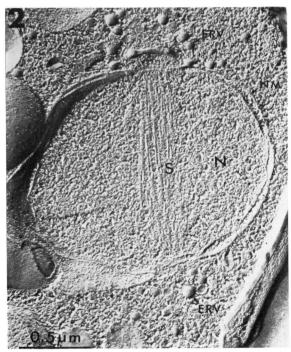

Figure 1.15 Freeze-etch electron micrograph of the spindle apparatus (S) in *Saccharomyces cerevisiae* in an early stage of meiosis during sporulation. N; nucleus; ERV: endoplasmic reticulum vesicles; NM: nuclear membrane. (Courtesy of Dr. S. F. Conti and the *Journal of Bacteriology*)

27

C

chromosome, e.g. chromosome 2 in *Neurospora*. Other RNA is present in the nucleus. This is heterogeneous in size and is presumed to be in part the precursor to the messenger RNA (mRNA) template for protein synthesis in the cytoplasm. Much of this heterogenous nuclear RNA is however degraded in the nucleus.

Segregation of chromosomes during mitosis and meiosis takes place on the *spindle apparatus*. The structure found during mitosis in *Saccharomyces* is depicted in Fig. 1.15. It is composed of a set of *microtubules* (filaments built up by assembly of subunit protein molecules in much the same way as bacterial flagella) emanating from a spindle plaque structure located just outside the nuclear membrane. In other eukaryotes the spindle plaque is replaced by a more conventional centriole structure.

Mitochondria

Mitochondria are found in most, if not all, microbial eukaryotes. These organelles are characteristic in electron micrographs, being bounded by two unit membranes. The inner membrane is convoluted into folds, called *cristae*, which are covered on their internal surface by granular structures. These contain ATPase activity and are the complexes involved in the oxidative metabolism of substrates and generation of ATP (see p. 67 and p. 83). Mitochondria are therefore specialised membranous structures in eukaryotes carrying out one of the functions of the bacterial cell membrane.

Mitochondria are needed only in microbial eukaryotes which metabolise aerobically. In some organisms capable of anaerobic growth the mitochondrion is repressed under anaerobic conditions or when glucose is present in excess. Such organisms, notably the yeast *Saccharomyces cerevisiae*, are very useful for studying the interactions between the organelle, the nucleus and the cytoplasm. The *petite* mutations in yeast lead to cells which have completely lost mitochondrial activity. These strains are viable provided they are supplied with a fermentable carbon source.

Mitochondria contain DNA capable of coding for some, but not all of their proteins. This DNA is usually circular, but is linear in *Tetrahymena.* They also contain a protein-synthesising system distinct from that of the eukaryotic cytoplasm. The mitochondrial system is inhibited by chloramphenicol, erythromycin and other inhibitors of bacterial protein synthesis, but not by cycloheximide which is specific to eukaryotic ribosomes. These inhibition data have been used to support the theory that mitochondria may be highly evolved bacterial endosymbionts, initially taken up by a plastid-free, amoeboid or phagocytic cell that depended on fermentation for its energy production. There are a number of important features found in mitochondria normally absent from bacteria, but one bacterial species so far examined, *Paracoccus denitrificans*, does possess many of these, including: phosphatidyl choline as a major component of membrane phospholipid; straight chain saturated and unsaturated fatty acids accounting for nearly all the membrane fatty acids; and a mitochondria-like respiratory chain (Fig. 4.6). It is interesting to note in this context that the giant amoeba *Pelomyxa*

28

palustris lacks mitochondria but contains endosymbiotic bacteria which presumably grow on lactic acid produced by the host fermentation of glucose. Similarly *Paramecium* species contain different 'species' of endosymbiotic bacteria which are associated with killer activity and are incapable of autonomous growth. It should be emphasised that there are other theories for the origin of mitochondria and chloroplasts; a recent one suggests that eukaryotes have evolved from phagocytic blue-green algae.

Chloroplasts

These are the eukaryotic organelles containing the photosynthetic pigments and enzymes needed for photosynthesis. Those from different groups of algae differ in photosynthetic pigments but have the same general structure. This involves bundles of closely stacked lamellae (thylakoids) each made up of a pair of membranes; the sets of lamellae are enclosed in a double unit membrane. The photosynthetic pigments (chlorophylls and carotenoids) are present in the lipid matrix of the thylakoid membranes.

Chloroplasts also contain DNA, and a protein-synthesising system which is sensitive to inhibitors of bacterial protein synthesis.

Ribosomes

Ribosomes form the granular background contrast in uranyl acetate-stained thin sections. Their structure and role in protein synthesis is discussed in Chapter 6. In eukaryotes there are two types of ribosomes: the majority are cytoplasmic and are either free in the cytoplasm or attached to regions of the endoplasmic reticulum; the others are located in the mitochondria and chloroplasts if they occur. Cytoplasmic ribosomes differ markedly from mitochondrial and bacterial forms in both size ($80\,S$ against $70\,S$) and in sensitivity to antibiotics. Cytoplasmic protein synthesis is not inhibited by chloramphenicol, erythromycin and tetracyclines, but is sensitive to cycloheximide which does not affect bacterial or mitochondrial ribosomes.

Careful extraction under the right ionic conditions yields *polysomes*, these are sets of ribosomes arranged linearly along mRNA. Each ribosome is associated with nascent polypeptide chains, and in bacteria the polysomes are formed as mRNA is transcribed from DNA (see Fig. 6.8).

Vacuoles

In some prokaryotes specialised gas-filled vacuoles enable the cells to adjust their buoyancy. These gas vacuoles are found mainly in purple photosynthetic bacteria and blue-green algae, and more rarely in other aquatic non-photosynthetic bacteria. They can occupy up to 40% of the cell volume, and are collapsible when exposed to sudden increase in pressure. They are surrounded by an unusual proteinaceous membrane which has a structure distinct from that of unit membranes.

In fungi and many algae large membrane-bound vacuoles are often found, particularly in aged cultures. Vacuoles isolated from yeast contain highly active lytic enzymes, including ribonuclease, esterases and proteases; their role may be to separate these enzymes from sensitive substrates within the cytoplasm. They may also act as sites for the storage or accumulation of ions and metabolites.

Protozoa contain a number of different vacuoles. During the phagocytosis of particulate substrates the plasma membrane pinches off to form a *food vacuole*, into which lytic enzymes are released to digest the food. Ciliates (the most familiar example is *Paramecium*) have an oral region leading into the cell to the site at which food vacuoles are formed. A second type of vacuole found in ciliates (and euglenid algae resembling protozoa) is the *contractile vacuole* which functions as a pump in media of low osmolarity to excrete water from the cell.

Inclusion Granules

In addition to ribosomes and organelles, the microbial cytoplasm often contains other particulate components whose occurrence varies with species and cultural conditions. Many of these 'inclusion granules' are storage compounds formed during growth in the presence of excess nutrients, these include: lipids, glycogen, starch, volutin (polymetaphosphate) and in some specialised bacteria (photosynthetic purple sulphur bacteria) molecular sulphur can accumulate as large intracellular crystals. These energy storage compounds are discussed in Chapter 4.

In the Gram-positive bacterial genera *Bacillus, Clostridium* and *Sporosarcina*, as well as most fungi and many algae, specialised resting stages or spores are produced within cells facing starvation. These specialised cells are discussed separately in the chapter on morphogenesis. The *parasporal crystals* associated with sporulation in *Bacillus thuringiensis* and several other *Bacillus* species are particularly interesting; each crystal is composed of protein antigenically related to spore coat protein, and is lethal to certain *Lepidoptera* species. These bacteria or their crystals are used in the biological control of several insect pests.

Apart from essential features, such as cell walls, cell membranes, nuclei, ribosomes and (in eukaryotes) mitochondria — no single type of prokaryotic or eukaryotic cell possesses all the structures outlined in this chapter. Some structures such as flagella or glycogen granules are widespread in their occurrence. Others, such as pili, have only been identified in a limited number of closely related bacterial species; generalisation is even more difficult in eukaryotes.

2 Growth and death

In this chapter we are concerned with dynamic aspects of microbial physiology: how microorganisms grow, what influences their growth, how they reproduce, how they respond to a hostile environment and how they ultimately die. Why are we interested in studying these phenomena? Obviously there are many practical reasons; on the one hand microbial growth can be deleterious and it is important to be able to control growth as much as possible. On the other hand growth of many microorganisms is used to produce or preserve human and animal food, to synthesise antibiotics and other palliatives such as ethanol, vitamins, essential amino acids and enzymes, or in treating or detoxifying waste materials. In addition to these practical interests, there is the fascination of trying to understand how cells reproduce.

What is Growth?

The answer to this question depends largely on one's point of view. In terms of a single cell, growth can be seen as an increase in size of the cell with time. Once this cell has reached a particular size it usually divides, and therefore growth can also be considered in terms of an increase in cell numbers. Thus the microbiologist may on the one hand be interested in growth of *individual* cells, including the way they increase in size, how they replicate essential structures and finally divide, or on the other, how a *population* increases in numbers or in total mass. These are two quite different aspects of growth and will be discussed separately.

GROWTH OF INDIVIDUAL CELLS

Cell Cycle

The term cell cycle has been coined for the process whereby a newborn cell grows by increase in size and ultimately undergoes division to form two cells. There are a number of ways in which microorganisms grow and divide; some are outlined in Fig. 2.1. Most bacteria, algae and protozoa divide by binary fission, each mother cell increasing in volume and eventually dividing to give two identical daughters. In some micoorganisms fission is asymmetric, and two non-identical cells result. A good example is the bacterial genus *Caulobacter* in which stalk cells divide to form motile cells and stalk cells. Another example is seen with division of diatoms,

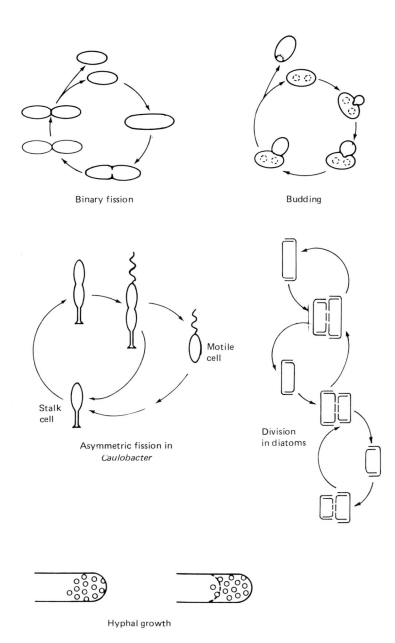

Binary fission

Budding

Motile cell

Stalk cell

Asymmetric fission in
Caulobacter

Division
in diatoms

Hyphal growth

Figure 2.1   Modes of cell growth and division

where the rigid silica walls (frustules) decrease in size at each successive division. Ultimately some of the daughters are so small that they undergo meiosis and form an auxospore; this resumes a larger size on germination.

Most yeasts, with the exception of the fission yeast *Schizosaccharomyces*, divide by budding, the mother cell barely changing in volume during the division cycle. A bud emerges at some point on the cell surface and increases in size until it almost matches the mother cell. Mother and daughter cells can be distinguished in this system by the number of bud scars present on the cell surface. Most fungi have a hyphal habit in which growth is usually restricted to the apical tip. In aseptate fungi it is difficult to define a cell cycle since long multinucleate hyphae are formed. The difference between yeast budding and hyphal growth is not as great as it appears; both involve cell extension at a restricted site and there are fungi which can reversibly switch from a mycelial to a yeast phase. This usually occurs after an increase in temperature or CO_2 concentration. Many of these fungi are potential pathogens.

Another form of growth is shown by the true slime moulds, e.g. *Physarum polycephalum*. These eukaryotes are *acellular*, increasing in size during growth but without dividing. A large plasmodial mass is formed which moves by cytoplasmic streaming. These amoeboid masses can divide and fuse randomly; they are multinucleate and their *nuclei* divide synchronously.

Kinetics of Cell Growth

Direct microscopic observation has shown that individual cells of most microorganisms increase in mass linearly with time. Volume changes are also roughly linear. Most growing cultures of microorganisms are *asynchronous*, cells from all stages of the cell cycle being represented in the culture at any time. Theoretically there are twice as many young cells of age 0 as there are old cells of age T_d (mean cell age at division). In fact the observed age distribution of cells in a growing culture deviates from theory since some cells divide at an earlier age than others, as indicated in Fig. 2.2. This means that for synchronised populations synchrony will decay fairly rapidly on further growth of the cells due to the distribution of individual cell generation times.

Synchronisation Procedures

In order to obtain enough material to study biochemical changes taking place during the cell cycle it is often necessary to synchronise the division of cells within a population. Since microorganisms can adapt very rapidly to changes in their environment it is important that methods used to obtain population synchrony do not introduce artifacts. There are two general ways of obtaining synchrony, either by *induction* (e.g. alternate temperature changes, light pulses to photosynthetic organisms, or the use of metabolic inhibitors affecting essential functions in the cell cycle) or by *selection* (e.g. on the basis of size using zonal density gradient centrifugation, or filtration, or by elution of newly divided cells from immobilised

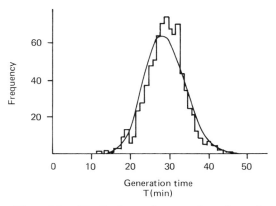

Figure 2.2 Distribution of generation times for exponentially growing *E. coli*. Histogram: experimental data; Curve: theoretical distribution fitted to data. From E. O. Powell and F. P. Errington, *Journal of General Microbiology* **31**, 315, 1963)

parents). Selection methods are less likely to introduce artifacts since the cells are selected on the basis of a physical characteristic (usually size) with the minimum of chemical disturbance; two methods in current use are outlined in Fig. 2.3.

Cell Cycle Events

There are many processes within a cell which are essential to its continued growth and division. Some of these are of a general metabolic nature (e.g. uptake of substrates, energy production, biosynthesis) which, while important to growth are not *specifically* concerned with maintaining the cell cycle. Here we are interested in those functions which are more intimately involved in the cell cycle. How can we determine what functions are relevant? In some cases, including the replication and segregation of DNA the answer is obvious, in others it can be more difficult to determine. Possibly the most useful criterion available is an inhibition test: if a cell function is specifically affected by either an inhibitor, or a mutation, does it cause an accumulation of cells *in a particular stage of the cell cycle*? If so then it is probably connected with the cell cycle in a fairly specific way.

Bacterial Cell Cycle

The major cytological changes seen during the bacterial cell cycle are cell growth, division by septum formation and DNA segregation into daughter cells. These, and the biochemical changes associated with gene expression have received most attention.

Growth of Surfaces This involves extension of cell walls and membranes. How does this occur — by random insertion of newly synthesised material or addition to

34

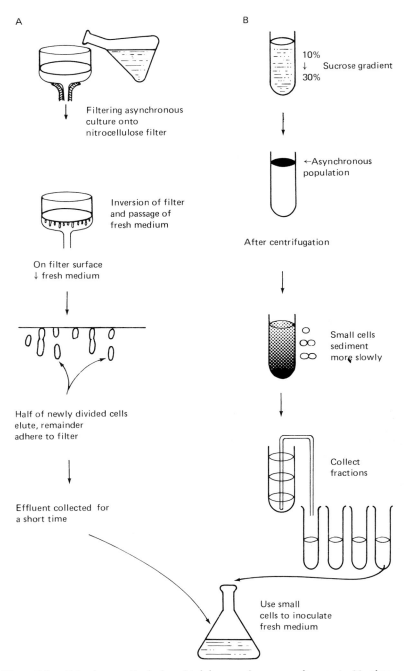

A

Filtering asynchronous
culture onto
nitrocellulose filter

Inversion of filter
and passage of
fresh medium

On filter surface
↓ fresh medium

Half of newly divided cells
elute, remainder
adhere to filter

Effluent collected for
a short time

B

10%
↓ Sucrose gradient
30%

←Asynchronous
population

After centrifugation

Small cells
sediment
more slowly

Collect
fractions

Use small
cells to inoculate
fresh medium

Figure 2.3 Selection methods for obtaining synchronous cultures; A, Membrane
elution B, Zonal centrifugation. The membrane elution method only works for those
organisms which attach to the membrane used

35

a few regions of the cell surface? For the cell wall it is quite clear that synthesis is restricted to particular regions of the preexisting wall. This has been shown by immunofluorescent staining techniques, using antibodies prepared against cell wall material, to distinguish new synthesis from old. In one organism, *Streptococcus faecalis*, electron microscopic studies have given a clear picture of the way new cell wall synthesis is organised. This represents a relatively simple situation, the division of a coccus in a single plane (Fig. 2.4). From this several points emerge:

i   Wall synthesis begins at an equatorial wall band formed during the previous division cycle, and continues at the leading edge of the nascent cross wall.

ii   The mesosome appears to play a role in organising the site of initiation of new cell wall synthesis, as well as attaching DNA to the membrane.

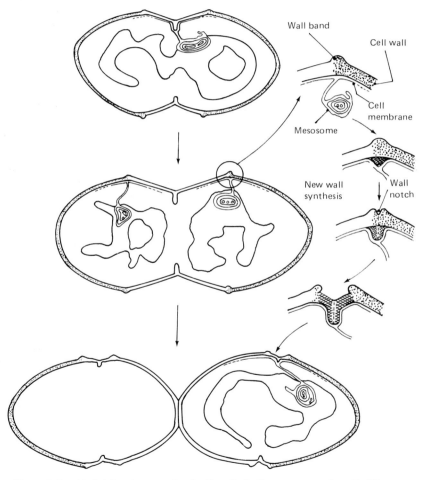

Figure 2.4   Model for the growth of cell walls in *Streptococcus faecalis*. (Courtesy of Dr. G. Shockman, reproduced with permission of the Chemical Rubber Company, from *Critical Reviews in Microbiology* **1**, 1, 1971)

iii   There is some indication in the figure, supported by much data from studies of the surface distribution of wall lytic enzymes, that growth of new cell wall depends on a limited breakdown of preexisting cell wall peptidoglycan. This is consistent with the observation that *autolysin*-deficient mutants of *Str. faecalis*, *E. coli* and *Bacillus subtilis* are unable to grow and divide properly unless lysozyme is added.

The mechanisms of wall synthesis and septum formation in bacilli and cocci dividing in more than one plane are less well understood, although similar principles of site-specific initiation and insertion of new cell wall, and of a requirement for lytic enzymes, apply. In bacilli there are probably two separate processes — lengthwise growth and septum formation.

Little is known about the synthesis of cell membrane, mainly since components of the membrane are semifluid and also undergo degradation and resynthesis (turnover) so that site-specific labelling is an uncertain proposition. There is some evidence that membrane synthesis is restricted to an equatorial band around the *B. subtilis* cell. When rod-shaped organisms divide there is an increase in the rate of incorporation of glycerol and phosphate into lipids, presumably reflecting the synthesis of the division septum.

DNA Replication   Mechanisms of DNA synthesis are discussed in Chapter 7, and here we are only concerned with how replication is coordinated with cell division. In the few bacteria studied in detail (*E. coli*, *Salmonella typhimurium*, *B. subtilis* and *Streptomyces coelicor*) there is one circular chromosome which, at least in *E. coli* and *B. subtilis* replicates bidirectionally from a single origin. This single chromosome takes a relatively constant time to replicate: about 40 minutes in *E. coli* at 37°C, at all except very slow growth rates. At high growth rates replication is too slow to duplicate the entire chromosome in one generation time, hence new replication forks are initiated at the origin before previous ones are completed. This means that under conditons of rapid growth there is more than one chromosome per cell; this is best understood by reference to Fig. 2.5.

Chromosome replication must be synchronised in some way with cell division if each daughter is to receive at least one copy of the genome. One important finding in *E. coli* is that the time from *the end of a round of DNA replication* (termination) *to the formation of a division septum* is also constant at all except slow growth rates. How is this timing achieved? So far the answer to this question is not complete, but Fig. 2.6 summarises one model which takes into account many of the data from mutant and inhibitor studies as well as the constant times for DNA replication and for the interval from the end of replication to septum formation. In addition it is based on the finding that the mass of a cell at the time of division (and its size) varies with growth rate, but that at the time of *initiation* of DNA replication the mass of a cell is constant regardless of growth rate. Important features of this model are:

i   A new round of chromosome replication is initiated at a particular cell mass, M_i, the 'initiation mass'. The initiation of replication depends on synthesis of protein at this time. DNA replication then takes a constant time regardless of growth rate.

Low growth rates (generation time greater than DNA replication time)

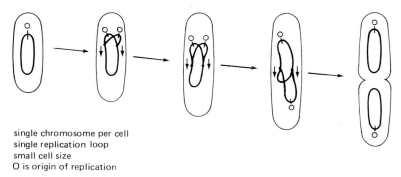

single chromosome per cell
single replication loop
small cell size
O is origin of replication

High growth rates (generation time less than DNA replication time)

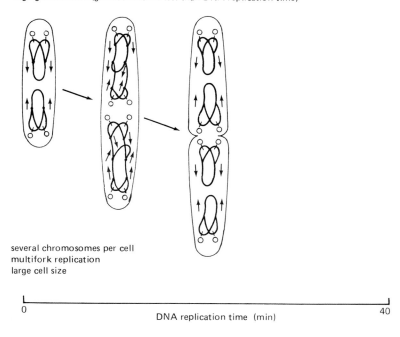

several chromosomes per cell
multifork replication
large cell size

|0| |40|

DNA replication time (min)

Figure 2.5 Replication of bacterial (*E. coli*) DNA. At low growth rates there is a single chromosome per cell and no reinitiation at the origin before completion of a round of replication. At higher growth rates there are multiple initiations of replication at the origin and more than one chromosome per cell.

ii A second sequence concerned with cell division is initiated at the same time. A period of protein synthesis is followed by another which may involve the assembly of a septum precursor.

iii When replication of DNA is complete a termination protein is synthesised and this interacts with the septum precursor leading to septum formation and cell division.

So far there is no clue as to what triggers DNA replication when a cell reaches a particular mass or size. Theories which have been proposed often include the concept of an inhibitor and/or initiator which is synthesised once per cell cycle and is diluted to a critical concentration by increase in cell volume.

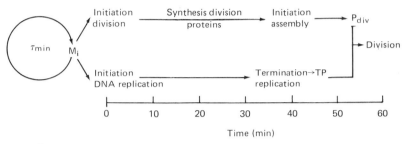

M_i is mass at initiation
τ is mass doubling time
TP is termination protein
P_{div} is division protein

Figure 2.6 Model for the coordination of DNA replication and cell division in *E. coli*. (Courtesy of Dr. W. Donachie and the Society for General Microbiology)

Gene Expression during the Cell Cycle This has been studied mainly in terms of the synthesis of those RNA species which are most easily identified (rRNA) or of the synthesis of particular proteins, mainly enzymes. Usually there is continuous synthesis of *total* protein and RNA during the bacterial cell cycle, but for individual enzymes the situation can be very different. If an enzyme were synthesised continuously throughout the cell cycle, in a synchronous culture, its activity would increase linearly with time until the gene coding for the enzyme doubled; then the rate should also double (provided the rate of enzyme synthesis is limited by the number of gene copies available for transcription). This is true for some enzymes, but many appear to be synthesised discontinuously, (Fig. 2.7), each one appearing at a characteristic time which reflects the order on the chromosome of the genes coding for these enzymes. The control mechanisms for sequential synthesis of enzymes remain a matter for speculation. One point however emerges very clearly: the composition of a microbial cell is in a considerable state of flux throughout its growth and division cycle underlining the considerable heterogeneity of asynchronous populations.

39

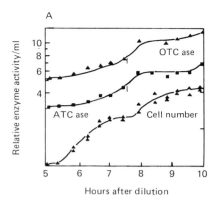

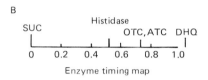

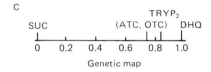

Figure 2.7 Step enzyme synthesis in synchronous cultures of *Bacillus subtilis*. A, The time course; B, a map of the approximate timing of enzyme synthesis during the cell cycle; C, The genetic map location of genes coding for synthesis of the enzymes. Note the correlation with the enzyme timing map. SUC: sucrase; ATC: aspartate transcarbamylase; OTC: ornithine transcarbamylase; DHQ: dehydroquinase. (Courtesy of Dr. M. Masters and National Academy of Sciences, U.S.A.).

Eukaryotic Cell Cycle

In many respects the patterns of control of the cell cycle in eukaryotes resemble those found in bacteria, with additional complexity due to the presence of organelles and the nuclear arrangement of DNA. One major difference is in the nature of DNA replication. In eukaryotes DNA replication occupies only a part of the cell cycle, which is therefore often split up into stages on this basis (G1 = first gap, S = DNA synthesis, G2 = second gap and M = mitosis). As an example, under certain conditions the timing of these phases in the ciliate *Tetrahymena pyriformis* is: G1, 50 min; S, 40 min; G2, 100 min and M, 30 min; this timing is very dependent on growth conditions. There is one similarity with bacterial systems: in yeast growing over a wide range of growth rates the S phase is relatively constant, and there is a roughly constant interval between the end of S and cell division.

40

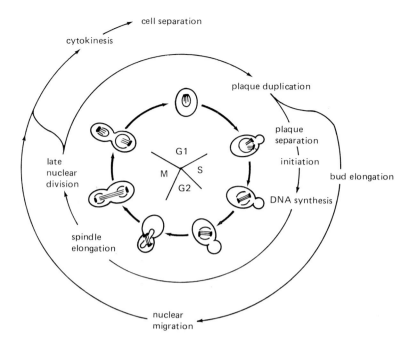

Figure 2.8   The cell cycle in *Saccharomyces cerevisiae*. Coordination of nuclear events and cell division. The spindle plaque replaces centrioles (see p. 28). (Redrawn with permission from Dr. L. L. Hartwell and the American Society for Microbiology)

The major events associated with the nucleus during the cell cycle of *Saccharomyces cerevisiae* are illustrated in Fig. 2.8, and superimposed on the diagram is the control circuit coordinating DNA replication with the cell division cycle. The main features of this circuit are very similar to those found in bacteria, e.g.:

i   There are two cycles. After plaque duplication, at the onset of mitosis, there are events associated with DNA replication and nuclear division which follow each other in sequence. The second cycle is concerned with morphological events: bud emergence and nuclear migration. Either of these two sequences can be blocked without affecting the other.

ii   The two sequences are coordinated since they merge at the stage of *cytokinesis* (the separation of cytoplasm preceding cell division), only the inner pathway need be completed in a cell cycle for a second to commence.

iii   Each is a *dependent* sequence of events, if one event is blocked by inhibition or mutation, no subsequent events on that sequence occur.

Organelle Duplication   Less is known about how organelles other than the nucleus grow and divide during the cell cycle, and how they are synchronised with nuclear or cell division.

41

Mitochondria and chloroplasts appear to divide by a fission process, seen most clearly in those small eukaryotes with a single representative organelle, e.g. the protozoan *Chromulina* and the alga *Chlorella*. In these the fission of the organelle usually occurs at or about the time of cell division. With microorganisms containing more than one mitochondrion division is difficult to follow in detail, although there is no evidence for synchronous fission. In yeast the mitochondria (from one to about ten per cell) are seen from serial sections to be sausage-shaped, usually one protrudes into the developing bud.

Replication of organelle DNA is not controlled in the same way as nuclear DNA since in a number of microbial eukaryotes examined, including *Tetrahymena, Chlamydomonas* and *Saccharomyces* the majority of the mitochondrial DNA is replicated at a different stage of the cell cycle from nuclear DNA. Some species of yeast are ideally suited to studying the control of mitochondrial DNA replication since they can survive if grown on glucose in the absence of the mitochondrion. Petites (small colonies lacking mitochondria) are induced by a very wide range of physical and chemical treatments, including DNA intercalating dyes, mutagens and inhibitors of cytoplasmic and mitochondrial protein synthesis. This may indicate that the coordination between the nucleus (which codes for many mitochondrial proteins) and the mitochondrion is particularly sensitive to disruption.

POPULATION GROWTH

The traditional method of studying microbial growth has been batch culture, in which a volume of medium is inoculated with a low number of organisms and their growth followed with time. This procedure has a number of disadvantages over other methods of growing cells since the culture medium is in a state of continuous change: it does however have the advantage of convenience.

Batch Culture

The nature of the growth curve obtained in batch cultivation is familiar; the various phases of growth and their significance are summarised in Table 2.1.

Exponential Growth

The rate at which the number, n, or mass, m, of microorganisms per ml increases in a culture depends on the number present at any time t.

i.e. $\quad \dfrac{\mathrm{d}n}{\mathrm{d}t} = \gamma n; \quad \dfrac{\mathrm{d}m}{\mathrm{d}t} = \mu m$

where γ and μ are *specific growth rate constants* for number or mass respectively. Under conditions of balanced growth γ is equivalent to μ. Integration gives the exponential function:

$$m_t = m_0 e^{\mu t}$$

m_t is mass/ml at time t, m_0 is mass/ml at zero time.

Table 2.1 Growth phases in batch culture

Growth phase	Cell activity
lag	Adaptation to new environment, synthesis of new enzymes. Also seen when large proportion of inoculum is composed of dead cells. Longer lag for starved cells or when fresh medium is not 'rich'.
log or exponential	Unhindered multiplication; all nutrients present in excess. Rate of growth determined by the composition of the medium and environmental factors.
late log	Concentration of one (or more) nutrients becomes limiting and falls to a low level with concomitant decline in growth rate.
stationary*	Can be due to nutrient deprivation or accumulation of inhibitory products of metabolism. Organisms still viable and using endogenous reserves for maintenance.
death	Occurs at a variable time after onset of stationary phase. Due to cell lysis by autolytic enzymes or effects of toxic metabolites.

*In some Gram-positive bacteria and many eukaryotes a process of sporulation is initiated at the end of log phase and is maintained by endogenous reserves accumulated during log phase.

The specific growth rate constant can readily be related to the mean generation time for the population, $\bar{\tau}$, by substituting $m_t/m_0 = 2$ and $t = \bar{\tau}$ in the exponential function:

$$\bar{\tau} = \frac{\ln 2}{\mu} = \frac{0.693}{\mu}$$

Either of these constants can be used in describing the rate of growth of a culture.

The growth rate of a culture is a very important parameter in microbiology, not only for its use in predicting concentrations of organisms at some future time, but also as a sensitive indicator of the *status* of a microorganism and its response to its environment. The influence of environmental factors on growth rate has been studied extensively, to develop and improve methods for preserving perishable commodities and ensure efficient product formation in fermentation processes. The factors affecting growth rate:

Availability of Nutrients

There are two facets to nutrient effects on growth rate; (i) concentration and (ii) quality. The effect on growth rate of altering the concentration of a particular nutrient *essential* for growth (in the presence of an excess of all other essential components) is illustrated in Fig. 2.9. This curve is hyperbolic and can be represented by the Monod equation:

$$\mu = \mu_m \left(\frac{s}{K_s + s} \right)$$

where μ_m = maximal growth rate obtained in excess nutrient, s = concentration of growth limiting substrate and K_s = half maximal saturation constant.

43

D

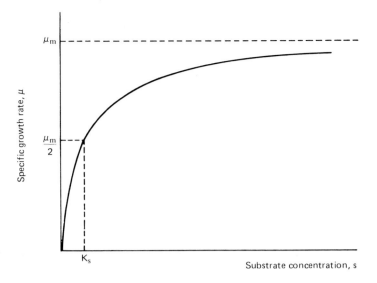

Figure 2.9 Variation of growth rate of a microorganism with the concentration of a limiting essential nutrient

This is analogous to the Michaelis-Menten equation for the rate of an enzyme-catalysed reaction. In general the values of K_s encountered for most growth substrates are very low, of the order 2 μM to 20 μM for sugars and 2 nM for amino acids, and reflect the high affinities of the systems for substrate uptake. This means that in a normal batch culture the specific growth rate μ approximates to μ_m until the end of the exponential phase of growth when very little of the growth-limiting substrate remains.

Quality of Nutrient Medium   Even in the presence of an excess of all medium components the growth rate of a microorganism can vary over a wide range, depending on the complexity of the medium. From Table 2.2 it can be seen that *Salmonella typhimurium* can grow on a simple medium containing one source of carbon and energy (e.g. lysine or glucose) and a mixture of inorganic salts which provides the N, P, S and trace element requirements. Some single carbon sources support a much higher specific growth rate than others, reflecting the extent to which each compound can provide energy for essential cell reactions and carbon metabolites for growth. On increasing the number of substrates available to a culture (e.g. by adding mixtures of amino acids) the growth rate increases further since the organisms use preformed precursors (of protein, etc.) and can redirect energy otherwise needed for precursor synthesis into processes essential to growth (e.g. solute uptake, synthesis of protein, RNA, membranes and cell walls, etc.) In the optimal case, obtained with brain-heart infusion broth, the mean generation time can be as low as 15 minutes.

Table 2.2 Growth rate as a function of medium composition, *Salmonella typhimurium*

Medium	Specific growth rate (μ, h^{-1})
lysine + salts	0.62
succinate + salts	0.94
glucose + salts	1.20
threonine + tyrosine + cysteine + histidine + phenylalanine + isoleucine + hydroxyproline + arginine	1.46
20 amino acids	1.83
casein amino acids	2.00
nutrient broth	2.75
brain-heart infusion broth	2.80

For laboratory purposes this method is commonly used to obtain different growth rates.

Temperature

This is one of the most important environmental variables affecting microbial growth. Every organism has a characteristic range of temperature over which it can grow. Most eukaryotic microorganisms do not grow and are killed above 45°C, although there are some fungi which can grow up to about 60°C, whereas bacterial growth is found between the extremes of − 7°C and 90°C.

For a particular organism there is a minimum, an optimum and a maximum temperature for growth, with the optimum fairly close to the maximum temperature. Organisms can be classified, according to their temperature dependence, into three groups. These are arbitrarily defined and do not in fact reflect the continuous nature of the microbial response to different temperatures; they do however provide useful categories in practice.

Psychrophiles   Definitions vary either to include those species capable of growth at 0°C or those with optimum temperatures below 20°C. Psychrophiles include many bacterial species, particularly *Pseudomonas*, *Flavobacterium*, *Achromobacter*, *Micrococcus* and *Serratia*, yeasts, other fungi and some algae. They grow, albeit slowly, at temperatures below 0°C.

Mesophiles   These have temperature optima around 37°C, and include most of the commonly encountered organisms.

Thermophiles   These grow at temperatures in excess of 50°C and include some fungi, blue-green algae, actinomycetes, and bacteria, particularly in the genera *Bacillus* and *Clostridium*. Thermophiles are found in almost any material in contact with soil.

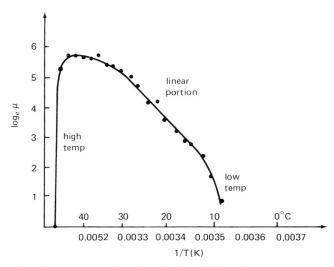

Figure 2.10 Arrhenius plot of the relationship between growth rate and temperature for a mesophilic microorganism (*E. coli*)

How does temperature affect growth rate? For a chemical reaction the rate varies with temperature according to the Arrhenius equation:

$$\log v = \frac{-\Delta H}{2.303\ RT} + \text{constant}$$

where v = rate, ΔH = activation energy, R = gas constant and T = absolute temperature. This gives a linear plot of *log rate* versus $1/T$. For most microorganisms the plot obtained resembles Fig. 2.10. At higher temperatures the nature of this curve is often explained in terms of competition between synthetic and degradative processes leading to a breakdown of cellular components (such as protein denaturation). At intermediate temperatures the Arrhenius relation applies, while at the lower end of the growth range there is a more rapid decrease in μ than expected from reaction rate theory. This could be due either to the breakdown of control mechanisms at low temperature (explicable if one considers the nature of allosteric enzymes, p. 150), or to changes in membrane fluidity due to crystallisation of some lipid components.

pH

As with temperature, every organism has a range of pH over which growth is possible, and an optimum pH. In general, bacteria are less tolerant of either pH extreme than fungi, but some bacteria can grow at lower pH values, including those producing acids such as lactic acid (*Lactobacillus* species), acetic acid (*Acetobacter*) and sulphuric acid (*Thiobacillus thiooxidans*). Most fungi have a minimum pH for growth between pH 1.5 and 2, but some species (*Acontium velatum* and

46

Dematiaceae species) have been reported to grow in 2.5 N H_2SO_4 saturated with $CuSO_4$.

Water Activity, a_w

In most microorganisms water is the major component of the cell (~80%), and is the solvent for many of the cellular reactions. There are a number of ways of effectively restricting the water available to the cell, including:

i   Dissolving salts or sugars to increase the osmotic pressure of the environment.
ii Physically removing water by evaporation or sublimation.
iii Freezing, which if done rapidly leads to crystallisation of ice within the microbe; slow freezing leads to the formation of ice crystals external to the cell.

None of these treatments is necesarily lethal to microorganisms.

The availability of water to an organism is indicated by *water activity*, a_w. This is the ratio of the vapour pressure of water in the solution under study to the vapour pressure of pure water at the same temperature and pressure:

$$a_w = \frac{p_{soln}}{p_{H_2O}}$$

The range of minimum water activities for growth of a number of microorganisms is given in Table 2.3. Most bacteria cannot grow below a_w of 0.90, with the exception of *Staphylococcus aureus* (0.86) and the *halophilic* bacteria (*Halobacterium* species) which can grow at a_w = 0.75 (equivalent to 5.5 M NaCl). Osmophilic yeast *Saccharomyces rouxii* and a number of xerophilic fungi can grow in high concentrations of sugars at an a_w of 0.6 (cf. saturated sucrose or 70% fructose has a_w = 0.76).

Growth Yield

Another important growth parameter is yield relative to a given substrate:

$$Y = \frac{\text{mass of bacteria formed}}{\text{mass of substrate used}} = \frac{m}{s}$$

The yield constant Y can be estimated in batch growth experiments. When cells enter stationary phase they have usually exhausted a compound essential for growth, the *limiting* component. Varying the initial concentration of this compound gives a plot of total growth versus substrate concentration which is usually linear and, from the slope, Y can be estimated. Y is a constant only for a particular organism under a given set of cultural conditions, and varying pH or temperature or any other parameter affecting the growth rate of the organism can lead to changes in the yield constant.

For heterotrophic organisms, the yield for carbon substrates is not directly related to the mass of carbon available to the cell, but depends on two factors. The

47

Table 2.3 Limiting water activities for growth of various microorganisms

Bacteria		Yeasts		Other fungi	
Bacillus mycoides	0.99	*Saccharomyces*	0.94	*Mucor spinosa*	0.93
Bacillus subtilis	0.95	*cerevisiae*		*Penicillium* species	0.8—0.9
Serratia marcescens	0.94	*Torula utilis*	0.94	*Aspergillus flavus*	0.90
Escherichia coli	0.93			*Aspergillus niger*	0.84
Sarcina species	0.92	**Osmophilic yeasts**			
		Saccharomyces	0.65	**Xerophilic fungi**	
Salt-tolerant bacteria		*rouxii*		*Xeromyces*	0.60
Staphylococcus aureus	0.87			*bisporus*	
Halophilic bacteria					
Halobacterium species	0.75				

first is the pathway by which the carbon substrate is metabolised to produce energy and metabolites needed for biosynthesis (discussed in Chapter 5). Secondly, when the carbon source on which yield constants are estimated is also the energy source, the yield is much more dependent on how efficiently the cell converts the chemical energy in the substrate to ATP, and how much ATP is expended in withstanding adverse effects of the environment.

Continuous Culture

In batch culture, microorganisms are subject to a continuously changing environment as substrates are consumed and products of metabolism accumulate, and exponential growth cannot be maintained indefinitely. There are, however, a number of ways of maintaining continuous growth and these are finding increasing application in research and industry. The advantages of continuous over batch culture can be summarised:

i   Smaller fermenter vessels are required and continuous operation reduces cleaning times and costs.
ii   Growth occurs at constant rate in a constant environment, and it is easier to control such parameters as temperature, pH and O_2 concentration at desired values.
iii   In some continuous systems growth rate can be varied in a controlled way enabling study of the effect of growth rate on cell size and composition.

The Chemostat

The simplest continuous arrangement, the chemostat, is depicted in Fig. 2.11. Fresh medium is pumped into the culture at constant flowrate, f and mixed as efficiently as possible. The culture volume, V, is maintained constant by a weir or syphon device so that excess culture is removed at the same rate as fresh medium flows in.

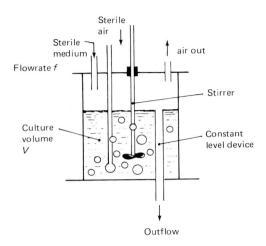

Figure 2.11   The chemostat

An important parameter in this system is the dilution rate, D, which indicates how fast the culture contents are diluted:

$$D = \frac{f}{V}$$

For a pure culture of an organism the rate of increase of cell mass in the culture vessel will be given by:

$$\frac{dm}{dt} = (\mu - D)m$$

We have seen previously that μ is a function of the concentration of the limiting substrate according to the Monod equation (p.43). If D exceeds μ_m, the maximal growth rate constant for the organism, washout of the culture will occur. If however $D < \mu_m$ the culture will adjust to a steady state, i.e. for $dm/dt = 0$:

$$\tilde{\mu} = D$$

where $\tilde{\mu}$ is steady-state growth rate.

This system is self-adjusting. From a low inoculum organisms will grow at μ_m and the population increase since there is an excess of substrate, s. As the population grows the rate of consumption of substrate increases, which for organisms in the culture is given by the yield equation:

$$-\frac{ds}{dt} = \frac{1}{Y}\frac{dm}{dt} = \frac{\mu m}{Y}$$

Eventually the population will consume the substrate at such a rate that the concentration in the vessel will decrease and with it the growth rate, μ, until a balance is reached.

49

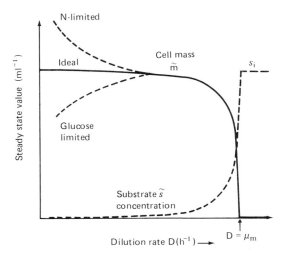

Figure 2.12 Theoretical curves for steady-state values of substrate concentration and cell concentration in a chemostat. Deviations from the ideal curve for cell concentration are indicated

The balance equation for substrate utilisation is given by:

$$\frac{ds}{dt} = \text{input} - \text{output} - \text{consumption by organisms}$$

$$= D(s_i - s) - \frac{\mu m}{Y}$$

where s_i is the concentration of the limiting substrate in the inflowing medium. At steady state $ds/dt = 0$, and therefore:

$$D(s_i - \tilde{s}) = \frac{\tilde{\mu}m}{Y}$$

By substituting for μ from the Monod equation and solving the two balance equations, it is possible to calculate the steady state values of substrate concentration, $\tilde{s}$, and cell mass, $\tilde{m}$ for any given dilution rate D in terms of the measurable constants K_s, μ_m, s_i and Y. The variation of $\tilde{m}$ and $\tilde{s}$ are plotted in Fig. 2.12.

$$\tilde{s} = K_s \left(\frac{D}{\mu_m - D} \right)$$

$$\tilde{m} = Y(s_i - \tilde{s})$$

At low growth rates under *glucose limitation, maintenance energy* (that is energy not used for growth and cell division, but to maintain existing cells in a viable state)

50

becomes significant. At low growth rates under *nitrogen source* limitation the opposite situation sometimes applies, and excess mass is obtained because glucose is diverted to production of large amounts of extracellular polysaccharide.

Applications and Limitations of Continuous Culture

Chemostat systems operate over a wide range of growth rates but provide conditions approximating to those found in a batch culture approaching stationary phase. If one is interested in a metabolite produced by a microorganism in full exponential growth or under strict starvation conditions a simple chemostat is inadequate, although multistage systems overcoming these limitations have been devised. Technical problems are also posed by organisms which adhere to surfaces or grow as filaments.

Mutation is a potential problem in a continuous system. For a neutral mutation such as resistance to bacteriophage which does not affect the kinetics of substrate utilisation, the mutant cells increase in a chemostat linearly with time. If, however, a mutant can use the limiting substrate more efficiently than the desired organism, it will rapidly take over the culture. For example, amino acid-overproducing strains are useful tools for producing biologically active (L)-amino acids by fermentation; these are however usually at a growth disadvantage with respect to non-over-producers which can easily arise by mutation. In some cases chemostats are useful in selecting mutants: by growing a continuous culture under limitation for a compound metabolised by inducible enzymes, *constitutive* mutants are rapidly selected (see Chapter 7). Moreover, they are very useful tools for studying physiological changes induced by varying the concentration of a single component in the growth medium. For example, bacterial sporulation, and the synthesis of many degradative enzymes are subject to repression by carbon and nitrogen sources, and chemostats are used in research on these phenomena. Multistage systems have also been used to study simple interactions between different microorganisms, as a first step towards analysing symbiotic, competitive and predatory interations in mixed populations.

SURVIVAL, INHIBITION AND DEATH

Of significant concern to many microbiologists are methods for either inhibiting or delaying growth, or for killing microorganisms. There are many physical and chemical treatments which can inhibit or kill microorganisms (Table 2.4). Each has an application in medicine, in preservation of perishable goods or in sterilisation processes. In each of these areas the range of methods available is limited, for example, in medicine the problem is one of *selective* destruction of microbes with minimal damage to the host; in food preservation, physical treatments alter flavour, texture and aroma and legislation restricts the use of chemical additives. The interest is therefore not only in methods for killing or inhibiting microorganisms, but also in the factors which influence survival or death, the organisms which are most resistant to treatments used, and how they survive exposure to stress.

Table 2.4 Effects on microorganisms of various extreme treatments, and the resistance and response to damage induced by the extreme condition

Extreme	Damage to cells	Resistant organisms	Adaptation or response to extreme condition
heat	enzyme denaturation	thermophiles	synthesis of more heat stable proteins
cold	loss of regulation; decreased membrane fluidity	psychrophiles	synthesis of higher proportion of unsaturated fatty acids
low a_w	dehydration and inhibition of enzyme activity	osmophiles, xerophiles, halophiles	accumulation of compensating solute or possession of enzymes adapted to high ionic strength
low pH	protein denaturation; inhibition of enzymes	acidophiles	active exclusion of protons, adaptation of surface appendages
ionising radiation	free radical or photon-induced damage to DNA and proteins	radiotolerant	enzymatic DNA repair

Response of Microorganisms to Environmental Extremes

There are a number of ways of looking at the response of an organism to adverse conditions; these include study of:

i   The physiology of those which show considerable resistance, including naturally resistant species (*Micrococcus radiodurans* to radiation; *Bacillus stearothermophilus* to heat) and resistant mutants of more sensitive organisms.

ii   Sensitive mutants, e.g. temperature-sensitive, radiation-sensitive

iii   Physiological changes in those exposed to near lethal extremes.

Usually there are many cellular processes which are impaired by a sublethal treatment and it is often difficult to determine which is most important to the survival of the cell. If we take as an example the effects of high temperature on bacteria, the following changes are induced: damage to the cytoplasmic membrane; breakdown of ribosomes; irreversible enzyme denaturation, and DNA strand breakage. Obviously any of these will lead to a reduction in the growth rate of a cell. What is less obvious is the extent to which each contributes to growth inhibition or death.

There are several ways by which an organism can survive and grow under conditions lethal to most others. These include:

Possession of more stable cell components   Many enzymes from the thermophile *Bacillus stearothermophilus* are less susceptible to heat denaturation than their counterparts in mesophilic *Bacillus* species. Halophilic bacteria have a cell membrane which is so adapted to high salt conditions that it dissolves at low ionic strength. These organisms avoid desiccation by being permeable to the salt in their environment, and their enzymes have been adapted to function at high ionic strength.

Effective repair systems   This applies particularly to radiation damage, there are at least three systems for repairing radiation-induced damage in *E. coli* and *Saccharomyces cerevisiae.*

Rapid resynthesis of damaged structures   One explanation sometimes given for the resistance of thermophiles to heat is that they can resynthesise protein at a rate which matches the rate of denaturation.

Compensation   One of the best examples is the way fungi and some salt-tolerant bacteria (*Staphylococcus* species) adapt to low a_w environments. Some fungi, including osmophilic yeast, synthesise large amounts of intracellular polyols, such as arabitol, to raise the internal osmotic pressure to that of the environment. In bacteria the accumulated solute is usually an amino acid, commonly proline.

Table 2.4 summarises the damage to, and response of, microorganisms to various environmental extremes. Some of these responses are general to all species, whereas others represent a specific evolutionary adaptation and are restricted to resistant species. Excluded from this table are bacterial spores, which are in general more resistant to physical and chemical extremes than vegetative cells. This resistance can be attributed to many factors, including the low permeability of dormant spores to non-aqueous solvents, modification of some enzymes during sporulation to increase heat resistance, stabilisation of structures by high concentrations of calcium dipicolinate (see Chapter 9), the arrangement of DNA in a different physical state from that found in vegetative bacteria, and the possession of highly efficient DNA repair enzymes activated after germination of the spore.

Death

Death of a microbial cell is usually defined as the loss of reproductive ability. Measuring death can present practical problems since the extent of survival of a population is very dependent on how the culture is treated after exposure to a lethal agent. This is particularly true of the medium used to detect survivors. In general richer media enable a higher degree of resuscitation. Survivors also show a very wide distribution in the times taken for recovery, some begin to divide many days after inoculating into a recovery medium.

Kinetics of Death

Cells in a population do not all die at once on exposure to lethal conditions. The kinetics of their death usually follows an exponential function as is illustrated by the curves obtained for *survivors* against *dose* for a particular treatment (Fig. 2.13).

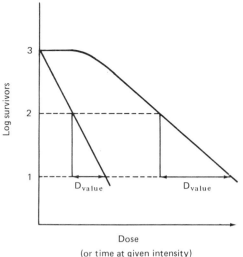

Figure 2.13 Two common examples of survivors versus dose curves for populations exposed to lethal agents

There is one important parameter obtained from dose/survivor curves. The dose (time at a given intensity) required to inactivate by ten-fold is the D value and is a measure of the resistance of an organism to a particular treatment.

Exponential loss of viability is often explained in terms of target theory, most easily understood for radiation-induced death. In every cell there is a function (or functions) sensitive to the lethal agent. For a given dose (a given number of bullets) the probability of a cell surviving depends on the proportion of targets which are not hit by these randomly sprayed 'bullets'. An approximate mathematical treatment of this simple model leads to an exponential function. A shoulder on the survival curve is often seen at low doses, particularly for radiation-induced damage, and is interpreted in terms of either the need for multiple hits, or the existence of systems for repairing damage after the lethal treatment.

This model is probably an oversimplification, since cells within a population have a range of susceptibilities and there are likely to be a number of different targets, each with different susceptibilities to a given dose. It does however focus attention on what might constitute target sites within cells. For damage induced by ionising radiation the primary target is quite clearly seen as DNA since mutants lacking enzymes involved in DNA repair are much more sensitive to various forms of radiation. For heat treatment, however, we have seen that there are many more candidates for targets since many functions are damaged by heat. There is however fairly strong evidence that death is ultimately the result of irreversible damage to membranes, since:

i   Leakage of cell constituents through membranes occurs after heat injury.

ii Mutants with temperature-sensitive lesions or an auxotrophic requrement can

54

usually recover after return to permissive growth conditions. Those which lose viability at the non-permissive condition are defective in membrane synthesis or function. It is interesting to note that these 'suicide' mutants can be prevented from losing viability at the restrictive condition by inhibiting protein synthesis. Death in these mutants (and possibly in other organisms) therefore appears to be the result of unbalanced growth due to defective membrane function and a loss of coordination between membrane and macromolecular synthesis.

Antimicrobial Agents

Antibiotics are compounds synthesised by microorganisms which kill or inhibit the growth of other microbial species, whereas antimetabolites are synthetic chemical agents active against microorganisms. For an antimicrobial agent to be useful in therapy it must have: *selective toxicity* against the pathogen and not the host; solubility in aqueous systems; and the ability to penetrate from sites of application to sites of infection. There are many compounds with antimicrobial activity, but relatively few of them meet these criteria. This is particularly true for antibiotics active against eukaryotic pathogens (e.g. fungi and amoebae), since in these there are fewer sites for selective toxicity than in bacteria with their different ribosomal composition.

Antibiotics, including many which have no clinical application, can be very useful laboratory tools since they are often specific inhibitors of particular metabolic functions.

Antimetabolites

Ehrlich, from studies on histological stains, conceived the idea of finding a dye with selective affinity for a parasite and attaching to it an active group lethal to the microorganism. In this way he hoped to obtain selectivity against pathogens. The first compounds tried were organoarsenic compounds, and eventually one, salvarsan, was found to be active against the causative agent of syphilis, *Treponema pallidum*.

The most successful antimicrobials however are the *sulphonamides*, all derivatives of sulphanilamide:

These compounds are bacteriostatic in their action, the effect *in vitro* being an inhibition of growth directly related to drug concentration. This inhibition is reversed in a competitive manner by the addition of p-aminobenzoic acid (PAB), a growth factor for some bacterial species. PAB is a component of folates, which are central to one-carbon metabolism, and sulphonamides inhibit the first enzymic stage in folate synthesis. Thus these drugs are active both against microorganisms

55

which can synthesise PAB and those which must be supplied with PAB for growth, but not against man who obtains folate from his diet.

Antibiotics

Antibiotics are synthesised by a wide range of microorganisms, many from *Streptomyces* species. Penicillin and cephalosporin have fungal origin while bacitracin and polymyxin are from *Bacillus* species. There are many poorly characterised antibiotics recorded in the literature; rather than list large numbers of these we will consider those best known in terms of their site of action, and illustrate a few cases of the mechanism of action. Structures of some of these are indicated in Fig. 2.14.

Compounds affecting cell wall synthesis   *Penicillin* inhibits prokaryotic cell wall synthesis causing an accumulation of cell wall precursors including the 'Park Nucleotides' (p. 132), and a further effect, which may in some species be the primary activity, is to prevent crosslinking of peptides on adjacent peptidoglycan chains (see p. 15). *Cycloserine* is a structural analogue of D-alanine (Fig. 2.14) and competitively inhibits both D-alanine formation and the subsequent synthesis of D-alanyl-D-alanine (see p. 133). Completion of peptidoglycan crosslinks is therefore inhibited and the resultant bacteria are as osmotically fragile as those treated with penicillin. *Vancomycin*, like penicillin, causes accumulation of peptidoglycan precursors and probably inhibits transfer of lipid-bound intermediate to the peptidoglycan acceptor, whereas *bacitracin* prevents dephosphorylation of the isoprenoid lipid carrier involved in peptidoglycan synthesis. All of these antibiotics lead to the formation of incomplete and structurally weak peptidoglycan or of no peptidoglycan at all and the cells therefore tend to lyse either through osmotic pressure differences or through the action of autolytic enzymes.

Antibiotics active against cell membranes   Several antibiotics bind to cytoplasmic membranes disrupting the normal membrane structure so that selective uptake of nutrients is prevented and cytoplasmic contents leak from the cell. Several antibiotics in this group are produced by sporulating bacilli; tyrocidin and gramicidin by *Bacillus brevis* and polymyxin by *Bacillus polymyxa.*

Tyrocidins and gramicidin are decapeptides consisting primarily of L-amino acids. They belong to a group of antibiotics called *ionophores* since they lead to an increased permeability of bacterial and eukaryotic membranes to ions, usually K^+, Na^+ or H^+ (see p. 66). Their mode of action can best be understood by considering their structure, outlined for gramicidin A in Fig. 2.14. This molecule forms a π helix which is left-handed and lipophilic, the amino acid side chains extend radially from the axis of the helix and become embedded in lipid of the membrane. The C=O groups are arranged in such a way that hydrogen-bonded head-to-head antibiotic dimer formation can occur; in the membrane this is sufficiently long to form a channel or hole across the membrane lipid bilayer, and explains the leakage of ions.

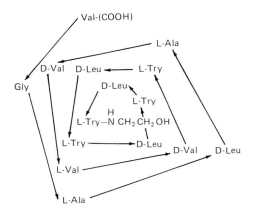

Val-(COOH)

L-Ala

D-Val D-Leu ◄─── L-Try

Gly

D-Leu

L-Try

H
L-Try—N CH₂ CH₂ OH

L-Try ────► D-Leu

D-Val   D-Leu

L-Val

L-Ala

Gramicidin A viewed
along axis of peptide
from C to N terminal end

D-*cycloserine*

D-*alanine*

puromycin

aminoacyl-t RNA

Figure 2.14   Some common antibiotics. Where the antibiotic is an analogue of a
cell metabolite that compound is also shown

Polymyxins are decapeptides in which the last seven residues form a cyclic heptapeptide and a branched chain fatty acid (6-methylheptanoate or 6-methyloctanoate) is linked to the first residue. This structure has a cationic detergent property which may account for some of its lethal effect on sensitive cells, although cationic detergents (including quaternary ammonium compounds) can also lead to precipitation of acidic molecules in the cell.

The *polyenes*, produced by some *Streptomyces* species, act only on eukaryotic membranes. They form complexes with membrane sterols and cause leakage of ions, solutes and even proteins. Some polyenes have specific ionophoric activity and uncouple oxidative phosphorylation in mitochondria. This group includes *nystatin* which is used to treat many fungal infections.

Antibiotics active against DNA, RNA and protein synthesis. There are many antibiotics of diverse structure and function included in this group, probably more than in any other. Some of these are summarised in Table 2.5 together with their probable site of action.

Many are inhibitors of bacterial protein synthesis, selectivity is due in large part to the differences between eukaryotic and prokaryotic ribosomes. Ribosomes perform a number of distinct operations during protein synthesis, and inhibitors for several of these are available. Other possible targets for selective inhibition of bacteria include DNA-dependent RNA polymerase and DNA polymerases; these enzymes are quite different between prokaryotic and eukaryotic systems (see Chapter 7).

For one of these antibiotics the mechanism of action is clearly understood. *Puromycin* inhibits both prokaryotic and eukaryotic protein synthesis since it mimics the structure of aminoacyl-tRNAs (see Fig. 2.14) and is incorporated into growing polypeptide chains. Since it does not have a free carboxyl group to accept another amino acyl group puromycin causes premature termination of polypeptide chain synthesis.

Resistance to Antimicrobial Agents

The use of many antimicrobial agents is becoming restricted to some extent by the emergence of resistant mutants, which in part is due to widespread abuse in the application of antibiotics. Resistance to a number of commonly used antibiotics, including penicillin, erythromycin and tetracyclines can be coded on extrachromosomal plasmids, *resistance transfer factors*, which are transferable from one bacterial species to another.

There are a number of ways cells can become resistant:

i   Bypassing the metabolic step affected.
ii   Overproducing an inhibited enzyme.
iii   Altering the structure of a target enzyme or other protein so that it can still function but no longer recognise the inhibitor. This is commonly the form of resistance to antibiotics affecting ribosomal functions, and in the extreme case, with streptomycin, it is possible to isolate strains which are *dependent* on the antibiotic for growth.

Table 2.5   Inhibitors of macromolecular synthesis

Synthesis	Inhibitor	Effective against	Site of inhibition
Protein	chloramphenicol	prokaryotes and mitochondria	inhibits peptide bond formation
	erythromycin	prokaryotes and mitochondria	inhibits translocation
	tetracyclines	prokaryotes and mitochondria	inhibits mRNA binding by the 30S ribosomal subunit
	streptomycin	prokaryotes and mitochondria	causes misreading of mRNA
	puromycin	prokaryotes and eukaryotes	premature termination of the growing peptide chain
	cycloheximide	eukaryotes, cytoplasmic only	affects 60S subunit
	cryptopleurine	eukaryotes, cytoplasmic only	affects 40S ribosomal subunit inhibiting translocation
RNA	rifamycins	prokaryotes	β-subunit of RNA polymerase
	lomofungin	eukaryotes	nuclear RNA synthesis, Mg^{2+} chelator
	α-amanitin	eukaryotes	specific to RNA polymerase synthesising mRNA
	actinomycin D	prokaryotes and eukaryotes	binds to DNA to inhibit transcription
DNA	nalidixic acid	prokaryotes	replication
	mitomycin C	prokaryotes and eukaryotes	crosslinks DNA
	hydroxyurea	eukaryotes	inhibits ribonucleotide reductase
	protein synthesis inhibitors	prokaryotes and eukaryotes	prevents initiation of replication

iv   By alteration of the membrane or its uptake systems so that the cell is no longer permeable to the drug. This is commonly seen when the drug is a competitive inhibitor of a low molecular weight metabolite (e.g. resistance to canavanine, an analogue of arginine, is usually acquired by the mutational loss of an arginine permease) (Chapter 3).

Resistant mutants are very useful laboratory tools. Where the site of action of an antibiotic is known, mutations to resistance provide a way of specifically affecting the function concerned. An example of the usefulness of this approach is the study of ribosome function using inhibitors of protein synthesis. Often mutation for resistance to an antibiotic affects a particular ribosomal protein, and with a knowledge of the step in protein synthesis inhibited by that antibiotic it is possible to identify which ribosomal subunit is involved in that step.

3 Substrate entry to the cell

Growth is a very dynamic process requiring considerable amounts of energy and substrates for component synthesis. Since the source of these substrates is the external environment the first essential steps in growth of a microorganism are locating substrates in the environment and transporting them into the cell. These processes in themselves require the expenditure of some energy by the cell, and this energy is derived from a coupling of intracellular metabolism to the transport processes.

Chemotaxis

In their natural environments most microbes are surrounded by ions and nutrients at very dilute concentrations; it is possibly an advantage to a cell if it can locate specific nutrients in great amounts. *Chemotaxis,* the ability to move along a concentration gradient of a compound in the supporting medium, has been shown to occur in a *few* motile species; attractants include some amino acids and sugars. The mechanism of this response is not understood, but it seems that a number of *chemoreceptors* (proteins specific to the attractant and possibly also involved in its transport) are involved. There are at least nine such receptors for sugars in *E. coli.* Chemotaxis in bacteria may be studied by inserting a capillary containing a solution of the attractant into a suspension of motile bacteria. Under positive chemotaxis the bacteria enter the capillary and can be assayed, while negative chemotaxis away from several amino acids, salts, acids, bases, alcohol and in some cases O_2 is also known to occur as indicated in Fig. 3.1.

UTILISATION OF HIGH MOLECULAR WEIGHT SUBSTRATES

Some microorganisms have adapted to efficient use of particulate substrates by producing structures for attaching their cells to surfaces. These include holdfasts, pili or fimbriae (in prokaryotes only) and stalks as in *Caulobacter.* Some eukaryotic microorganisms, including ciliate and amoeboid protozoa and slime moulds, ingest particles into membrane-bound food vacuoles prior to their digestion to smaller molecular weight compounds. Others, those surrounded by rigid cell walls, cannot ingest particulate matter directly.

There are then two initial mechanisms used by microbes in dealing with larger substrates: *engulfment* and subsequent degradation by intracellular enzymes, and *secretion* of enzymes to hydrolyse them into smaller molecules. These two processes are not as different as they may at first appear since engulfment occurs

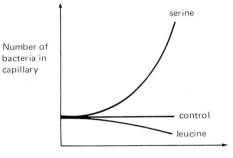

serine

Number of
bacteria in
capillary

control

leucine

Log concentration of compound in capillary

Figure 3.1 Positive and negative chemotaxis
in *E. coli.* Serine is an attractant; leucine is a
repellant.

into membrane-bound vacuoles; in both processes there is the same problem of transporting enzymes across membranes. Engulfment is the more efficient process since the substrates and enzymes needed to break them down are confined within a vacuole and neither enzymes nor degradation products are lost to the cell.

In microorganisms with rigid walls hydrolytic enzymes are secreted into the culture medium. Enzymes found in the supernatant after removing cells from a culture by centrifugation can have different origins:

i   Intracellular enzymes released from the cytoplasm of dead or leaking cells.

Table 3.1   Some microbial extracellular enzymes

Microorganisms	Enzymes	Mol. wt	Substrates
	Depolymerases		
Bacillus subtilis	α-amylase	45 000	starch
	proteases		proteins
Pseudomonas saccharophila	α-amylase	40 000	starch
Trichoderma viride	cellulase		cellulose
Cytophaga species	agarase		agar
Vibrio cholerae	neuraminidase	20 000	neuraminic acid
Streptococcus haemolyticus	DNase		DNA
	RNase		RNA
Clostridium welchii	proteases		proteins, polypeptides
Bacillus megaterium	proteases		proteins, polypeptides
Micrococcus lysodeikticus	proteases		proteins, polypeptides
	Other enzymes		
Clostridium welchii	lipase		lipids
Staphylococcus aureus	phosphodiesterase		esters
Bacillus cereus	penicillinase		penicillin
Staphylococcus aureus	penicillinase		penicillin

61

ii Extracellular ones which are synthesised at and extruded through the cell membrane in prokaryotes, or into vesicles which fuse with the membrane in eukaryotes. In Gram-positive bacteria these appear in the culture fluid, whereas in Gram-negatives some, such as ribonuclease and alkaline phosphatase, are retained in the *periplasm* (the region between the outer and cytoplasmic membranes). There are many bacterial examples of extracellular enzymes hydrolysing a wide variety of polymers and other substrates including nucleic acids, proteins, lipids and carbohydrates (including the cell walls of other bacteria) (Table 3.1). Myxobacteria are particularly active, producing a battery of enzymes to lyse and degrade other bacteria in their environment. Eukaryotes also produce extracellular enzymes.

In most microorganisms the synthesis of extracellular enzymes is tightly regulated. Low molecular weight substrates are usually preferred; extracellular enzymes are not usually synthesised until the cells face starvation (for example the amylases, ribonuclease and proteases in *Bacillus* species). The mechanism of this control is not fully understood, but it probably is analogous to catabolite repression of the synthesis of intracellular enzymes (p. 147). In some cases an extracellular enzyme breaks a polymer down into oligomers which act as inducers of further specific hydrolases needed to complete hydrolysis to compounds which can be taken up by the cell.

Entry of Small Molecular Weight Nutrients: Permeation

At this point we should recall the structure of the cytoplasmic membrane (p. 18); a lipid bilayer containing proteins embedded in, on, or across the hydrophobic matrix. Polar or ionic species cannot therefore pass freely across the membrane, but lipid-soluble substrates such as glycerol can enter the cell by free diffusion. This process is known as *passive transport*. The rate of entry of these compounds into the cell depends on the difference in concentration between the inside and outside of the membrane, and such a solute will only enter the cell if its external concentration exceeds that inside. For passive entry the rate of uptake of related compounds is greatly affected by the size and charge of the molecule concerned.

Active Transport

The absorption of most potential substrates, and many ions, relies on active transport complexes located in the membrane. These are so named because they can couple transport to an energy-yielding process, and internal solute concentrations in excess of those outside the cell are usually obtained.

A great deal of research has been carried out on investigation of transport across membranes, most of it concerned with ion uptake by mitochondria, although some progress has been made in finding out how solutes are transported across the bacterial membrane. Before discussing how this energy-coupled transport occurs we should look at the major characteristics of these systems.

62

Table 3.2 Specificity of the *Escherichia coli* β-galactoside permease

Substrate	K_m *binding of substrate (mM)*	*% Displacement of TDG from complex with permease protein*
lactose	0.07	—
thiomethylgalactoside	0.5	1
thiodigalactoside (TDG)	0.02	(100)
melibiose	0.2	95
α-methylglucoside	—	0
galactose	—	0
glucose	—	0

Specificity　Transport systems can show a very high degree of specificity. This can be shown by using structural analogues of a compound or mutants defective in its uptake. The most widely studied example is that for uptake of β-galactosides (including lactose) in *E. coli*, and the specificity of this system for some β-galactosides is indicated in Table 3.2. In other cases, such as transport of amino acids, there is a lower degree of specificity. L-leucine, L-isoleucine and L-valine share a common uptake system in *E. coli* as do the aromatic amino acids. This is also the case in the eukaryote *S. cerevisiae*.

Involvement of membrane proteins　Specificity is largely determined by a protein component of the system. These proteins are called *permeases* since they show an enzyme-like affinity for the molecules they transport. Permease synthesis is often regulated in the same way as the synthesis of enzymes involved in the metabolism of the substrate once it has entered the cell, (see p. 138). The M-protein, the *E. coli* permease for lactose and other β-galactosides, has been isolated from the membrane using detergent treatments; the difficulty in extracting this protein has been taken as an indication of its firm attachment to the membrane matrix. Permeases for many other systems have been identified by biochemical characterisation of mutants which are unable to metabolise a particular solute despite having a complete set within the cell of the enzymes for its degradation. These are *cryptic* mutants. Sometimes permeability mutants can be selected by using inhibitors. L-canavanine is a toxic analogue of L-arginine and acts as a substrate of the permease in *S. cerevisiae*. Mutants resistant to this analogue are defective in this permease. It should be understood that a permease is one component in a multicomponent transport system, although the term is often used loosely to denote the whole complex.

Energetics　Active transport systems can concentrate some solutes 100 to 1000-fold over the level in the external environment. This can lead to internal cell concentrations of 0.1 to 0.2 M for some sugars. The expenditure of considerable amounts of energy is obviously needed. How this energy is coupled into membrane transport is the subject of considerable debate, although there is much more agreement on its source. For the uptake of ions (e.g. H^+ and Ca^{2+}) by mitochondria, and ions and *some* solutes by respiring bacteria, the energy is derived from the oxidation of substrates via the respiratory chain (p. 81). Some inhibitors of the respiratory chain, in particular uncouplers of oxidative phosphorylation e.g. dinitrophenol, inhibit transport. In other cases, particularly in those organisms growing anaerobically without a respiratory chain (e.g. *Streptococcus faecalis*) other energy sources are important.

MECHANISM OF TRANSPORT

Sugar transport in bacteria is particularly interesting and illustrates a number of principles. Sugars occur within the cell almost exclusively as phosphorylated derivatives, and phosphorylation occurs during transport by the sequence:

(i) phosphoenolpyruvate (PEP) + protein $\xrightarrow[Mg^{2+}]{enzyme\ I}$

$$\text{phosphorylprotein + pyruvate}$$

(ii) phosphorylprotein + sugar $\xrightarrow{enzyme\ II}$ sugar phosphate + protein

Step (i) involves the substrate-level phosphorylation of the histidyl group of a low molecular weight soluble protein, catalysed by a soluble enzyme (I). The phosphorylated protein can transfer its phosphoryl group to any of a number of sugars, including in *E. coli*, glucose, mannose, fructose, mannitol and β-glucosides. This second step is catalysed by a family of membrane-bound enzymes, each sugar specific. The energy is derived from the hydrolysis of energy-rich phosphoenolpyruvate. Not all sugars are transported in this way since β-galactoside uptake is respiration linked, but in *Staphylococcus aureus* it represents the major mode of sugar uptake.

Not all compounds are chemically modified during transport; most are not, but there is some evidence that purines, pyrimidines and fatty acids may also be phosphorylated.

Respiration-linked Transport

The model proposed by Kepes for *respiration*-linked transport is given in Fig. 3.2. In this model, a permease molecule recognises and binds the substrate which it then transports across the membrane. This may be due to a conformation change induced in the permease by binding of the substrate. The upper half of this diagram represents the situation when the substrate is not accumulating against a concentration gradient since it has been found for uptake of β-galactosides that energy is not expended under these conditions. This type of uptake is called *facilitated diffusion*. When necessary the process is coupled into an energy-dependent reaction which alters the affinity of the permease for the substrate on the inner side of the membrane.

64

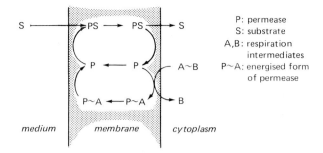

Figure 3.2   Model for the β-galactoside permease activity of *E. coli* suggested by Kepes

The precise mechanism for this coupling is not known, but several theories have been put forward which have been the subject of considerable debate. These theories include the following:

Mitchell's Chemiosmotic Theory   The respiratory chain is oriented across the membrane (which must be impermeable to H^+ and OH^- ions) such that oxidation of a substrate leads to translocation of protons from the inner side of the membrane to the outer, see Fig. 3.3. This generates an ion gradient creating a chemical and electrical potential capable of driving a reaction requiring energy. For example, the uptake of Ca^{2+} ions by mitochondria is driven by the membrane potential generated by H^+ translocation. On the other hand phosphate ion transport is thought to occur via a dual transport of H^+ and HPO_4^{2-} and is driven by the pH gradient established by respiration. This theory requires that membranes be intact for all reactions coupled to respiration (including oxidative phosphorylation discussed in the next Chapter).

Chemical Coupling   The theories in this group all suggest that energy coupling is mediated by a chemical reaction involving an energy-rich intermediate formed during respiration. The transport of sugars by the phosphorylation system described above is due to a direct chemical coupling of transport to the hydrolysis of phosphoenolpyruvate.

So far there is no definitive evidence to settle the debate, although the current trend is to favour the chemiosmotic theory since:

i   Reactions coupled to respiration (e.g. oxidative phosphorylation) do not occur in membranes or vesicles which are not intact (closed).

ii   The mitochondrial and bacterial membranes are inherently impermeable to ions, including H^+ and OH^-, and all transport systems so far examined function so as to maintain osmotic, pH and ionic equilibrium.

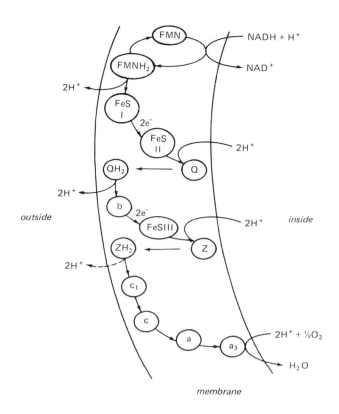

Q = coenzyme Q
Z = hypothetical hydrogen carrier
FeS = nonhaem iron proteins
b,c,a,a_3, cytochromes

Figure 3.3 Chemi-osmotic theory of respiration-linked transport.
Redrawn from Harold, *Bacteriological Reviews*, **36**, 177 (1972)

iii   A number of antibiotics or antimetabolites affect membrane function by allowing the specific conduction of H^+ or K^+ ions across the membrane. These *ionophores* uncouple oxidative phosphorylation and inhibit active transport. Included in the ionophores are the group of antibiotics known as the macrolides (valinomycin, gramicidin, nigericin) which are large molecules forming a hydrophobic shell around a central 'pore' region. In the pore region ions can form clathrate compounds; for some macrolides this is specific to one ion, others are less selective.

Solute Transport in the Mitochondrion

Transport of solutes into the mitochondrion does not involve direct coupling to respiration. For malate, the uptake system involves a specific exchange at the

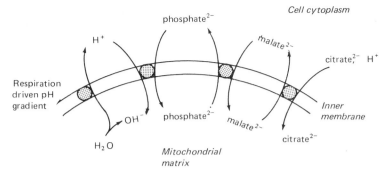

Figure 3.4   Solute uptake by mitochondria

membrane of phosphate for malate, energy being put into this system by the previous accumulation of phosphate ions. For citrate there is an even less direct link since citrate is specifically exchanged for malate (Fig. 3.4). The prime mover in all solute uptake therefore appears to be the generation of a pH gradient across the membrane by respiration.

Transport in Bacterial Vesicles

In order to study transport across the bacterial membrane it is necessary to have an intact closed structure. The bacterial cell is such a closed form but is much too complicated to use in studying the interaction of components during transport. Recently considerable progress has been made by the development of methods for preparing closed membrane vesicles from bacteria. Spheroplasts can be lysed under very carefully controlled conditions to give vesicles with the original membrane

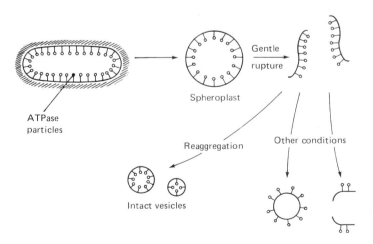

Figure 3.5   Preparation of membrane vesicles from bacteria

67

polarity preserved (ATPase stalks inward, Fig. 3.5) and with very little contamination by cytoplasmic enzymes or cofactors. These vesicles can transport ions and solutes against a concentration gradient provided a suitable energy source is added. Vesicles have been prepared from both Gram-positive and Gram-negative bacteria including *E. coli, Bacillus subtilis, Staphylococcus aureus* and *Mycobacterium phlei.* They have been very useful tools since in them transport is still coupled to respiration, and by using different oxidation substrates it is possible to explore the site of energy coupling in the respiratory chain (see p. 81) during respiration-linked uptake.

4 Energy production

All living systems need to expend energy, not only for growth, but also to survive when not growing. They need it to do mechanical work during propulsion by flagella, during cytoplasmic streaming, or when separating chromosomes during cell or nuclear division. Energy is needed for chemical work to drive highly endergonic reactions such as those concerned in the synthesis of macromolecules and their subunit monomers from the substrates available to the cell. They need to do work to maintain osmotic differentials, and to transport ions and molecules across membranes against concentration gradients. A few are even capable of coupling their energy producing systems to the emission of light.

Where does this energy come from? Ultimately it is derived from the energy of sunlight, either directly by transfer of photon energy to chemical bond energy as in photosynthetic bacteria and algae, or through chemical transformation of organic molecules derived from photosynthesis. There is one exception, in that *chemolithotrophic* bacteria can obtain energy from reactions involving inorganic compounds. The manner in which energy is obtained from chemical sources varies with the type of microorganism; some processes are limited to a very few species, while others are of widespread occurrence. Usually exotic substrates are metabolised by a restricted number of organisms to intermediates which are then metabolised by enzyme systems common to many organisms.

Nutritional Categories

On the basis of energy source, microbes can be subdivided into two major groups. Those using light directly are *phototrophs*, while those oxidising either organic or inorganic compounds are termed *chemotrophs*. Also there is subdivision on the nature of the carbon source: in *lithotrophs* the carbon requirements for synthesis of cell components can be satisfied by reduction of CO_2; in *organotrophs* organic compounds provide the main source of carbon. All microorganisms can therefore be subdivided into four major groups on the basis of their modes of nutrition and sources of energy (Table 4.1).

Energy Coupling

The majority of reactions producing energy are oxidations. How is the energy released in these reactions coupled to do useful work? If an oxidation-reduction

Table 4.1 Nutritional categories of microorganisms

Nutritional type	Principal energy source	Principal carbon source	Found in
Photolithotrophic	Light	CO_2	Plants, many algae, some bacteria
Photoorganotrophic	Light	Organic compounds	Algae, bacteria
Chemolithotrophic	Oxidation of inorganic compounds	CO_2	Bacteria
Chemoorganotrophic	Oxidation of organic compounds	Organic compounds	Animals, fungi, protozoa, many bacteria

couple occurs with direct electron transfer then the energy produced is dissipated as heat and is lost to the organism. If the two half-cell reactions are separated, but connected electrically so that electron or ion transfer occurs through a circuit, then useful work can be done. This type of coupling can be achieved in the cell in a number of ways:

i   *Direct chemical coupling* to the synthesis of an energy-rich bond can occur. The example illustrated is an important reaction from the fermentation pathway (see p. 72) and is an example of the *substrate-level* phosphorylation of ADP.

glyceraldehyde-   1,3-diphosphoglycerate   3-phosphoglycerate
3-phosphate

Notice that the oxidation of glyceraldehyde-3-phosphate to 3-phosphoglycerate occurs *via* an intermediate which enables the phosphorylation of ADP to ATP. Some of the free energy change in this oxidation-reduction reaction is conserved in the ATP phosphoryl bond.

ii   *Electrical coupling*   We have already discussed in a previous chapter (see Fig. 3.3) how a series of redox reactions can be coupled into proton or ion translocation across membranes. The potential difference or pH gradient formed can be used to transport ions, or as we shall see later (p. 83) to synthesise energy-rich bonds by reversing energetically-unfavourable reactions such as the phosphorylation of ADP.

ATP and related high energy molecules are formed during catabolic reactions that involve the oxidation of organic substrates (or during photosynthesis in the coupled transport of electrons ejected from chlorophyll by absorption of photons). There are a number of ways they can provide energy for the endergonic reactions of biosynthesis. These can involve coupling in the opposite direction to that found in substrate-level phosphorylation, so that the intermediate is activated by phosphorylation. Another important mechanism seen commonly in the activation of monomers during polysaccharide synthesis is the formation of nucleotide diphospho intermediates: e.g. bacterial glycogen synthesis:

$$\text{glucose-1-phosphate} \xrightarrow{\text{ATP} \quad P_i} \text{ADP-glucose} \xrightarrow[\text{primer}]{\text{ADP}} \text{glycogen}$$

There are other possibilities, one system worth exploring further being protein synthesis in which ATP and GTP participate in activation and translocation reactions in a number of ways (see p. 126).

Coupling of Reducing Equivalents

Since the breakdown of energy-yielding substrates usually involves their oxidation, an acceptor of reducing equivalents is needed. On the other hand there are many biosynthetic processes which require reduction steps. In order to couple oxidations to reductions in a general way, several important *coenzymes* act as electron (and hydrogen) acceptors and donors. Thus although most enzymes are quite specific in the reactions that they catalyse, a common redox coupling mechanism allows one oxidation reaction to provide reducing equivalents to many other reactions.

$$\begin{array}{c} AH_2 \\ A \end{array} \bigg\rangle \begin{array}{c} NAD^+ \\ NADH \\ + H^+ \end{array} \bigg\langle \begin{array}{c} CH_2 \\ C \\ BH_2 \end{array} \quad B$$

The two most common coenzymes fulfilling this function are nicotinamide adenine dinucleotide (NAD^+) and its phosphorylated derivative $NADP^+$.

Moreover NADH can be reoxidised by the respiratory chain (p. 81) which is the major site of energy generation in many microbial systems and the cofactor therefore also acts as a common link between many oxidation reactions and the respiratory chain.

MAIN ENERGY PATHWAYS

Heterotrophic organisms can use a wide variety of carbon substrates as energy source. These include carbohydrates, amino acids, fatty acids, alkanes, purines and pyrimidines. In general a few substrates are preferred and these support the highest growth rates (see section on catabolite repression, p. 147). For many

71

microorganisms the preferred substrates are carbohydrates, usually glucose, although there are bacteria which cannot use glucose as a source of energy.

Two terms describe different types of energy-yielding systems. *Fermentation* has found very wide meaning in industrial terminology, but in its formal sense is used for those pathways in which organic compounds serve as both electron donors and electron acceptors. *Respiration* is applied when oxygen, or some other inorganic compound or ion acts as the terminal electron acceptor.

Fermentation

Fermentable substrates include carbohydrates, organic acids and amino acids. A variety of fermentative pathways are available to microorganisms.

Glucose metabolism has received considerable attention from biochemists not only because of energy requirements, but also since it contributes many building blocks for biosynthesis. There are three pathways known in microorganisms; these differ in the initial steps, but share common later reactions leading to the conversion of 3-carbon sugar phosphates to pyruvate. Pyruvate is not the end product of most fermentations, but is the common precursor to most.

Hexose Diphosphate Pathway (Embden-Meyerhof-Parnas Scheme)

Louis Pasteur first showed a balance between the amount of glucose fermented by yeast and the alcohol and carbon dioxide formed. The reaction sequence in yeast, outlined in Fig. 4.1, was the first major pathway to be elucidated by biochemists, and occurs in most microorganisms. The essential features of this pathway are:

i   Glucose is activated by phosphorylation (either during or after entry to the cell) in a step requiring the conversion of ATP to ADP. A second phosphorylation forming fructose-1,6-diphosphate uses a second ATP. There is however *net synthesis* of two ATP molecules per molecule of glucose oxidised since two 3-carbon sugar fragments are formed in the step catalysed by *aldolase*, and the conversion of each of these to pyruvate leads to phosphorylation of two ADP molecules.

ii   The phosphorylation steps involve substrate-level coupling of the energy of oxidation to ATP synthesis, at a thermodynamic efficiency of about 30%.

iii   During conversion of glucose to pyruvate, NAD^+ is reduced to NADH. Since NAD^+ is in limited supply in the cell the metabolism of glucose would cease unless the NADH so formed was reoxidised. There are many reactions found in different microorganisms for converting pyruvate to final fermentation products with the balanced reoxidation of NADH. The same organism may produce different compounds when grown in different environments, e.g. under anaerobic conditions or in glucose excess, yeast cytoplasmic enzymes ferment glucose to ethanol, but in the presence of bisulphite (forming an addition product with acetaldehyde) they convert glucose to glycerol by the NADH-linked reduction of glyceraldehyde-3-phosphate to glycerol-1-phosphate.

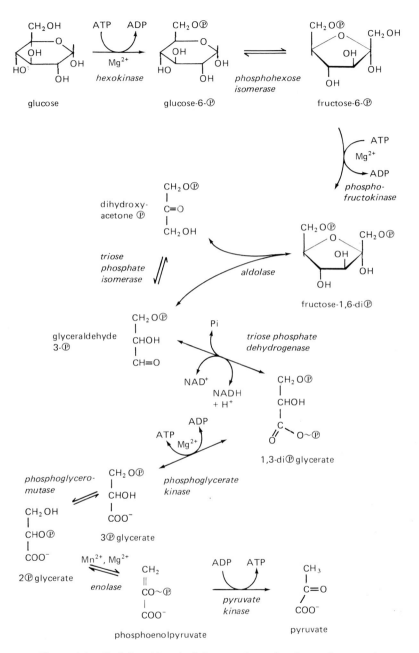

Figure 4.1 Embden—Meyerhof—Parnas scheme for glucose fermentation

In several bacteria, e.g. *Lactobacillus, Bacillus, Streptococcus* and *Clostridium*, the homolactic fermentation converts all the pyruvate to lactate.

$$CH_3COCOO^- + \xrightleftharpoons[\text{lactate dehydrogenase}]{\substack{NADH \quad NAD^+ \\ + H^+}} CH_3CHOHCOO^-$$

pyruvate $\qquad$ *lactate dehydrogenase* $\qquad$ lactate

Thus lactate is the major product of the fermentation of hexoses in these bacteria and the process can be represented by the equation:

$$C_6H_{12}O_6 + 2\ Pi + 2\ ADP \longrightarrow 2CH_3CHOHCOOH + 2\ ATP + 2\ H_2O$$

The enzymes involved are mainly soluble proteins found in the microbial cytoplasm or associated with the cytoplasmic membrane. Bacteria such as *Escherichia coli* and related facultative anaerobes perform a *mixed acid fermentation* which yields ethanol (in small amounts) along with other products:

$$2(C_6H_{12}O_6) + H_2O \longrightarrow C_2H_5OH + 2\ CH_3CHOHCOOH$$

hexose $\qquad\qquad\qquad$ ethanol $\qquad$ lactate

$$+ CH_3COOH + 2\ CO_2 + 2\ H_2$$

acetate

The difference in the end products in these microorganisms reflects the fate of pyruvate and not a fundamental difference in the hexose diphosphate pathway. Some enteric bacteria are anaerogenic, i.e. no gas is evolved from degradation of the carbohydrate substrate. These species, including *Shigella dysenteriae*, produce a 2-carbon fragment (as acetyl-CoA) and formate from pyruvic acid. The enzyme *hydrogenlyase* is absent from *Shigella* and the formate is not further degraded to CO_2 and hydrogen as in *E. coli*.

$$H^+ + HCOO^- \longrightarrow H_2 + CO_2$$

Another group of closely related bacteria, *Klebsiella, Serratia* and *Erwinia*, produce a neutral product, butane-2,3-diol. This characteristic product is used in the diagnostic differentiation of these genera.

$$2\ CH_3COCOO^- \longrightarrow CH_3\overset{\displaystyle COCH_3}{\underset{\displaystyle |}{C}}HCOO^- + CO_2$$

pyruvate $\qquad\qquad\qquad$ acetolactate $\qquad\downarrow$

$$CH_3CHOHCOCH_3 + CO_2$$

acetoin $\qquad\downarrow$

$$CH_3CHOHCHOHCH_3$$

2,3-butanediol

Further variations in the final products from the hexose diphosphate pathway are found in other microbial groups (see Table 4.2).

74

Table 4.2   Products derived from the hexose diphosphate pathway

Fermentation	Products	Microorganisms
Homolactic	Lactate	Streptococcus, Lactobacillus
Mixed acid	Lactate, acetate,* succinate, formate, ethanol	Escherichia, Proteus, Shigella, Salmonella
Butanediol	As in 'mixed acid', butanediol†	Klebsiella, Serratia, Erwinia
Butanol	Acetate, butyrate, butanol, acetone, ethanol, i-propanol	Clostridium species
Butyric acid	Acetate, butyrate*	Clostridium, Butyribacterium
Propionic acid	Acetate, propionate, succinate†	Propionibacterium, Veillonella

*Also CO_2 and H_2
†Also CO_2

Phosphoketolase Pathway (Warburg-Dickens pathway)

The hexose phosphate pathway does not enable microorganisms to form pentoses or to use them as a source of energy. In some bacteria, e.g. *Leuconostoc* species and some *Lactobacillus* species the phosphorylation of glucose is followed by its oxidation to 6-phosphogluconate. Further oxidation and decarboxylation of this intermediate yields a pentose phosphate which is split by the enzyme phosphoketolase to give a 2-carbon fragment (acetyl phosphate) and triose phosphate. Subsequently these products are converted to ethanol and lactate respectively; the overall reaction scheme is shown in Fig. 4.2, and can be represented by the equations:

$$\text{Glucose} \longrightarrow \text{Lactate} + \text{Ethanol} + CO_2 + \text{ATP}$$

and

$$\text{Ribose} \longrightarrow \text{Lactate} + \text{Acetate} + 2\text{ATP}$$

Organisms which carry out this fermentation lack aldolase needed to cleave hexose phosphates into two 3-carbon fragments. A characteristic of this pathway is that the yield of one ATP per molecule of pyruvate formed from glucose is half that obtained in the Embden-Meyerhof pathway. Another feature is that roughly equal amounts of lactic acid, ethanol and CO_2 are produced (other compounds such as glycerol and formate are formed in other reactions). This type of fermentation is often called the 'heterolactic fermentation' and is readily distinguishable from homolactic since CO_2 is not produced in the latter, and the ATP yield of heterofermenters for a given amount of glucose oxidised is less.

Entner-Doudoroff Pathway

Isotopic studies on *Zymomonas lindneri* showed that pyruvate could not have been formed from glucose by the reactions of the hexose diphosphate pathway, although

F

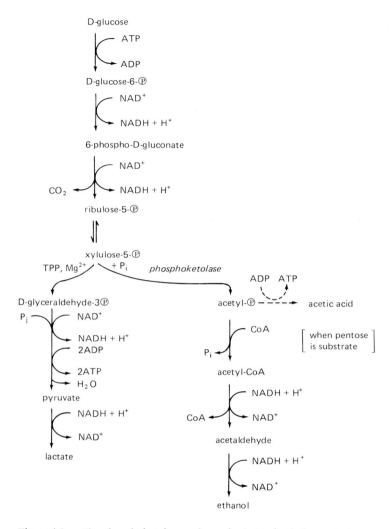

Figure 4.2   The phosphoketolase pathway for heterolactic fermentation

the initial reaction, the formation of D-glucose-6-phosphate is the same. The remaining reactions are shown in Fig. 4.3. The characteristics of this modified pathway are: (i) the formation of 3-deoxy-2-oxo-6-phosphogluconate; (ii) the labelling pattern in which the carboxyl group of one mole of pyruvate is derived from carbon atom 1 of the hexose and that of the second pyruvate molecule is formed from carbon atom 4 of the hexose; (iii) the lower energy yield of 1 mole ATP per mole of glucose – due to the oxidation of a single mole of triose phosphate as opposed to the two moles available in the hexose diphosphate pathway. The process is much less widely used by microorganisms than the hexose

76

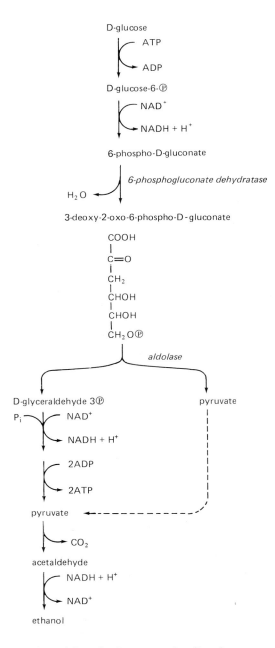

D-glucose

ATP

ADP

D-glucose-6-℗

NAD⁺

NADH + H⁺

6-phospho-D-gluconate

6-phosphogluconate dehydratase

H₂O

3-deoxy-2-oxo-6-phospho-D-gluconate

```
COOH
|
C=O
|
CH₂
|
CHOH
|
CHOH
|
CH₂O℗
```

aldolase

D-glyceraldehyde 3℗   pyruvate

Pᵢ   NAD⁺

NADH + H⁺

2ADP

2ATP

pyruvate

CO₂

acetaldehyde

NADH + H⁺

NAD⁺

ethanol

Figure 4.3   The Entner-Doudoroff pathway

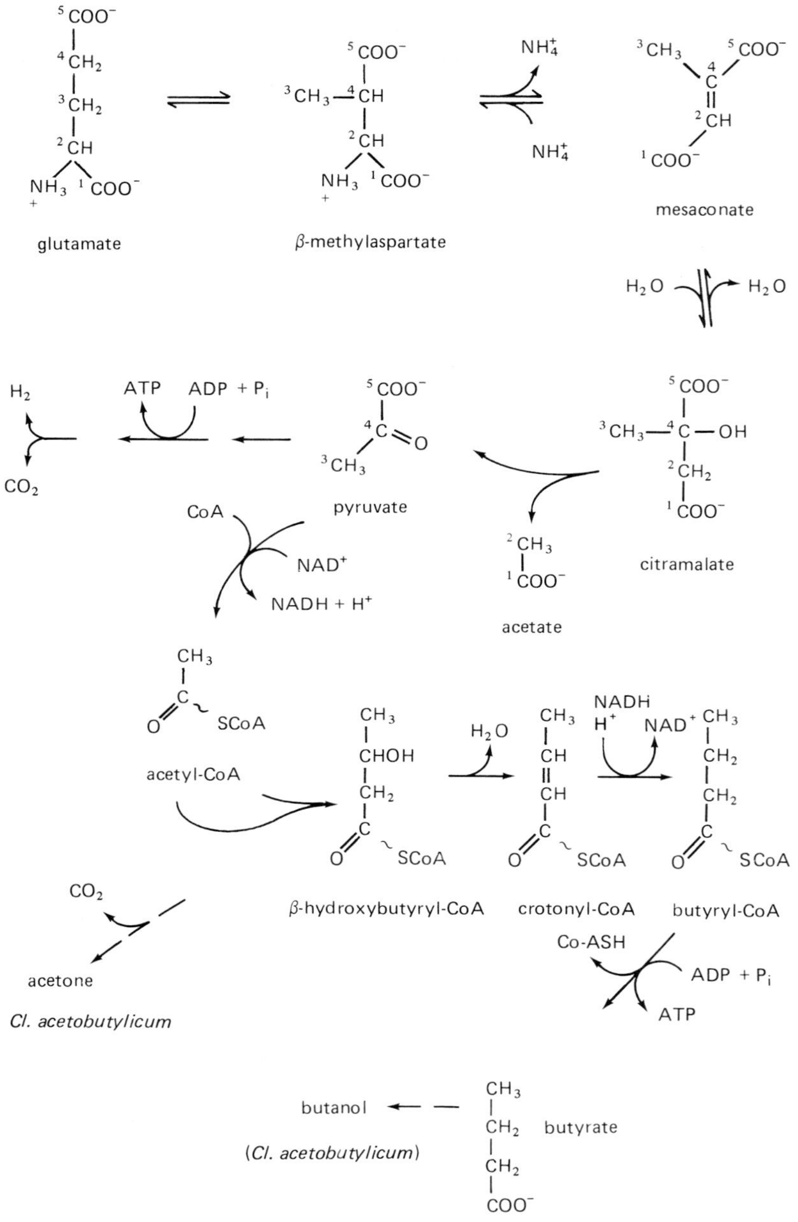

Figure 4.4 Fermentation of glutamate by *Clostridium* species

diphosphate pathway, but occurs in a number of *Pseudomonas* species and a few other Gram-negative bacteria. It can also be induced in Gram-positive *Streptococcus faecalis* by growth on gluconate. In *Z. lindneri*, the major products of glucose breakdown are ethanol and carbon dioxide, which are produced from pyruvate by the same mechanism as in *Saccharomyces cerevisiae*.

Fermentation of Nitrogenous Compounds

Fermentation of amino acids and other nitrogenous compounds is also important. The microorganisms mainly responsible are either anaerobic sporing rods (*Clostridium* species) or anaerobic cocci (*Micrococcus* species). The catabolism of glutamate by *Clostridium tetanomorphum* is typical; the rate resembles that of glucose breakdown by other microorganisms. The major products of glutamate fermentation are acetate, butyrate, hydrogen, ammonia and carbon dioxide, thus differing from the products obtained when glutamate is metabolised to α-ketoglutarate and the TCA cycle. The main steps in glutamate fermentation are shown in Fig. 4.4. From this it can be seen that:

i   An initial rearrangement of glutamate occurs. Isotopic labelling has shown this involves the carbon atom at the 3-position.
ii   Pyruvate and acetate are formed as cleavage products. The pyruvate can be oxidised or converted to a number of products in a series of transformations which provide energy, either as ATP or acyl-CoA derivatives.

Respiration

Fermentation is a relatively inefficient process since glucose is not completely oxidised, and much more energy is available if it can be converted to CO_2 and water. In *aerobic* respiration oxygen is ultimately reduced, but the coupling of this reaction to substrate oxidation is indirect. Electrons are transferred from substrates to NAD^+ and subsequently proceed via a series of intermediate carriers (the respiratory chain) to the step involving oxygen. Most of the energy obtained during respiration is derived from the oxidation of NADH by the respiratory chain and can be used to do work in solute and ion transport (p. 65) or can be coupled to the biosynthesis of ATP.

Respiration can therefore be studied from two angles, one dealing with the flow of carbon compounds to CO_2 and reduction of NAD^+, the other with electron transport and oxidative phosphorylation in the respiratory chain.

Tricarboxylic Acid Cycle (TCA or Krebs Cycle)

The principal intermediate of hexose metabolism during glycolysis is pyruvate; it is also derived from the oxidation of many amino acids and lipids. Under aerobic conditions pyruvate is initially converted to acetyl-CoA by an oxidative decarboxylation reaction sequence involving the participation of several cofactors and catalysed by a multienzyme complex.

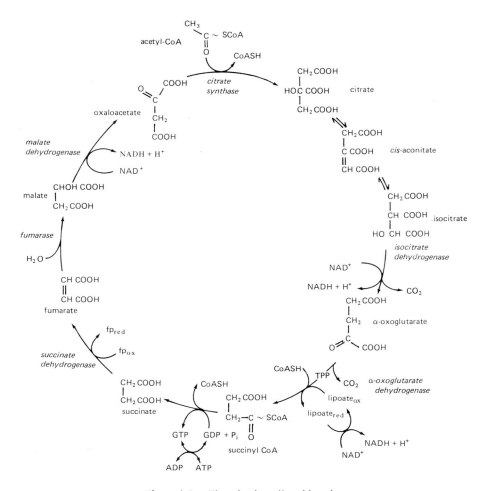

Figure 4.5 The tricarboxylic acid cycle

Acetyl-CoA is the substrate for the TCA cycle, which is set out in Fig. 4.5. The important features of this cycle can be summarised:

i   Acetyl-CoA condenses with oxaloacetate to form citrate. Completing the cycle once results in the oxidation of the acetyl group to two CO_2 molecules and the regeneration of oxaloacetate. Thus the intermediates of the TCA cycle are effectively *catalysts* and are not generated in substrate amounts.

ii   There are four oxidation-reduction steps. At three of these NAD^+ is reduced to NADH (in some microbial systems $NADP^+$ acts as a substrate for isocitrate dehydrogenase), while at the succinate to fumarate step the flavin prosthetic group (fp) of succinate dehydrogenase is reduced.

iii   The oxidative decarboxylation of α-ketoglutarate follows the same mechanism as the oxidation of pyruvate to acetyl-CoA. At the end of this sequence there is one substrate level phosphorylation of ATP coupled to the hydrolysis of succinyl-CoA.

80

Central Role of the TCA Cycle

The TCA cycle performs a number of functions other than the generation of energy. Some intermediates of the cycle are precursors in the biosynthesis of important metabolites including amino acids, purines and pyrimidines. They are also produced by breakdown of many metabolites, and the enzymes of the TCA cycle therefore assist in the interconversion of compounds such as the formation of aspartate from glutamate during sporulation in *Bacillus subtilis*. If, during these reactions, intermediates of the TCA cycle are drained off into biosynthesis then replenishment reactions known as *anaplerotic* sequences, must operate (see Chapter 5).

Electron Transfer, Respiratory Chain and Oxidative Phosphorylation

We have seen that NAD^+, lipoate and the flavin prosthetic group of succinate dehydrogenase are reduced during the fermentation of glucose to pyruvate and its oxidation to CO_2 via pyruvate decarboxylase and the TCA cycle. These steps are summarised in Table 4.3. Under aerobic conditions the electrons produced in the reoxidation of NADH and flavins are transferred to oxygen:

$$O_2 + 4H^+ + 4e^- \longrightarrow 2H_2O$$

This electron transfer occurs via a sequence of intermediate electron carriers in such a way that there are discrete steps of energy release, at three of these the free energy change is sufficiently great for coupling to the phosphorylation of ADP (Fig. 4.6). However, in bacterial systems the complete respiratory chain is not always present, and oxidative phosphorylation rarely yields 3 molecules of ATP per molecule NADH oxidised or atom of oxygen reduced.

The characteristic electron transport chain of eukaryotes is located in the mitochondrial inner membrane. In bacteria the respiratory chain is a component of the cell membrane and is in general terms similar to that found in the mitochondrion, but as indicated in Fig. 4.6 varies from species to species in the nature and number of components present.

Table 4.3 ATP yield in aerobic respiration via the TCA cycle

	ATP
Glucose + 2 NAD$^+$ $\longrightarrow$ 2 pyruvate + 2 NADH	8
pyruvate + NAD$^+$ $\longrightarrow$ acetyl–CoA + NADH	3
isocitrate + NAD$^+$ $\longrightarrow$ α-oxoglutarate + NADH	3
α-oxoglutarate + NAD$^+$ $\longrightarrow$ succinyl–CoA + NADH	3
succinyl–CoA $\longrightarrow$ succinate	1
succinate + flavin $\longrightarrow$ fumarate + flavin H$_2$	2
malate + NAD$^+$ $\longrightarrow$ oxaloacetate + NADH	3
pyruvate $\longrightarrow$ CO$_2$	15
glucose $\longrightarrow$ CO$_2$	38

The major components of the electron transport chain are flavoproteins, quinones and cytochromes. Flavoproteins contain a prosthetic group, either FAD or FMN, which is firmly bound to the protein and acts as acceptor of reducing equivalents from NADH. The quinonoid compounds can be of two types: *ubiquinones* are derivatives of 2,2-dimethoxy-5-methylbenzoquinone with isoprenoid chains 6 to 10 units long attached to carbon atom 6; the *vitamin K* group are derivatives of 2-methyl-1,4-naphthaquinone. Gram-positive bacteria

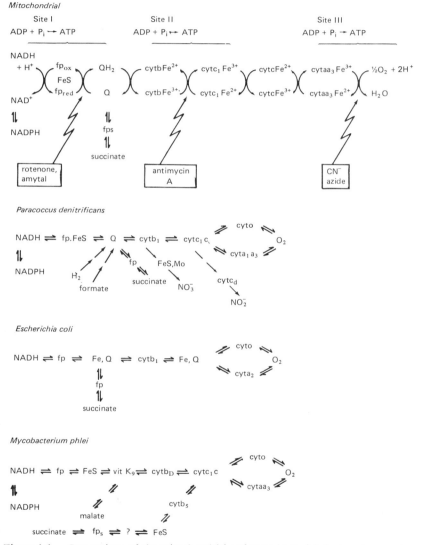

Figure 4.6 Comparison of the mitochondrial and some bacterial electron transport chains. For the mitochondrial chain the sites of coupled phosphorylation of ADP and of action of some inhibitors of respiration are shown

contain mainly naphthaquinones, facultatively anaerobic Gram-negative bacteria have both, while in others ubiquinones form the major constituents.

The cytochromes reveal the diversity of bacterial systems as opposed to those of other microorganisms. Identical cytochrome types are found in all eukaryotic mitochondria, whereas bacteria contain a wide variety of cytochromes, few of which correspond directly to those in mitochondria.

Energy coupling to ATP Formation

We have already discussed the coupling of energy released in electron transport to ion or solute transport and the theories proposed for the mechanism (p. 64). As with ion transport, oxidative phosphorylation is susceptible to uncouplers, including ionophorous antibiotics, and can only occur in intact preparations. Associated with all preparations capable of coupled ATP biosynthesis, and intimately associated with the process, is an ATPase activity. This activity is a complex of a number of polypeptides located in granules found on the inner surface of the mitochondrial inner membrane or of the bacterial cell membrane. The ATPase activity in mitochondria has been studied in some detail using a number of specific inhibitors and techniques for reversibly dissociating the complex. In eukaryotes, mutations conferring resistance to the inhibitors show mitochondrial inheritance, indicating that some components of the ATPase complex are coded by the mitochondrion. The synthesis of others is inhibited by cycloheximide (an inhibitor or cytoplasmic protein synthesis) and these are probably coded by the nucleus.

Chemiosmotic Hypothesis

According to this hypothesis the respiratory chain is organised across the membrane in such a way that electron transport causes excretion of protons. This generates across the membrane a pH and electrical gradient which can be used to reverse the ATPase reaction (Fig. 4.7). The ATPase complex functions so that proton movement inwards with the pH gradient drives the reaction to the synthesis of ATP. This can be visualised as an enzyme complex located in the membrane such that ATP, ADP, phosphate and H^+ have access to the active site from the inner side only and OH^- from the outside only. Synthesis of ATP would be favoured by a pH gradient since the condensation of ADP and phosphate to ATP would be favoured by the outward transfer of OH^- and the inward transfer of H^+ tending to neutralise the gradient.

Chemical Coupling Hypothesis

Direct chemical coupling theories draw an analogy with substrate-level phosphorylation reactions: one of the carriers of the respiratory chain interacts with a hypothetical intermediate which is activated. This intermediate could then participate in the phosphorylation of ADP. From what is known of the biochemistry of the electron transport chain this scheme must be slightly more complex than other substrate-level systems and requires the participation of at least

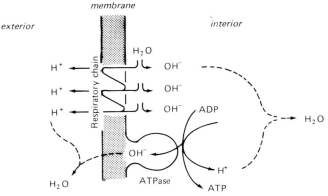

Figure 4.7 Diagrammatic representation of the coupled phosphorylation of ADP

two intermediates, I and X in the following scheme:

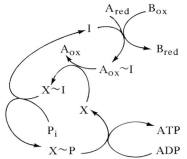

A_{red}, A_{ox}, B_{red}, B_{ox}: components of respiratory chain.
X,I: intermediates in oxidative phosphorylation.

There has been no completely satisfactory demonstration of the existence of coupling factors, and there is some vagueness as to the biochemical nature of the ATPase mechanism proposed by Mitchell's chemiosmotic hypothesis. Hence the current extensive interest in the nature and organisation of the ATPase complex.

Anaerobic Respiration

Some facultatively anaerobic bacteria use an oxidative type of metabolism under anaerobic conditions. This process is *anaerobic respiration*. The pathways for the catabolic degradation of the carbon and energy sources are identical with those in aerobic respiration, and electron transport occurs via a respiratory chain similar to that in aerobic cells. However, oxygen is replaced as terminal electron acceptor by another inorganic compound or ion, frequently nitrate. In this role metabolism of the inorganic coumpound is dissimilatory: nitrate is reduced to nitrite without incorporation of the inorganic nitrogen into cell components. Nitrite is toxic to most bacteria, but some can use it as an electron acceptor and the molecular N_2 formed is lost to the environment. In some species nitrate is successively reduced to nitrite, then nitrogen; this process of *denitrification* is found in several bacterial genera including *Pseudomonas*. Inorganic electron acceptors such as nitrate have a lower redox potential than oxygen, and the amount of ATP produced in systems with nitrate reductase is therefore usually less than in aerobic bacteria with cytochrome oxidase.

Fermentation versus Respiration

We have seen that for every molecule of glucose fermented there are two molecules of ATP formed. In aerobic respiration the ATP yield is considerably greater, as indicated in Table 4.3, 38 molecules of ATP are produced for every molecule of glucose completely oxidised. In terms of the total free energy available from the complete oxidation of glucose, about 30% is recovered in the aerobic pathway, the rest is released as heat. This heat must be removed in large scale industrial growth of microorganisms where cooling can be expensive.

Pentose Phosphate Cycle

This cycle is found in many bacteria and most eukaryotic organisms. It can serve several functions in cells.

i   There is conversion of 6-carbon sugars to others, including 5-carbon sugars needed for nucleic acid synthesis and 7-carbon sugars for the formation of aromatic amino acids.

ii   It is extremely important for autotrophic organisms since it enables products of carbon dioxide fixation to be converted to central metabolites.

iii   It serves as a source of NADPH supplying reducing equivalents to biosynthetic reactions.

iv   It can provide energy to the cell as an alternate pathway for the oxidation of glucose, and a mechanism for obtaining energy from pentoses.

The initial steps in this 'cycle' involve the oxidation of glucose-6-phosphate (Fig. 4.8).

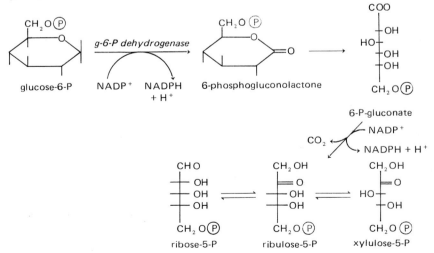

Figure 4.8   Oxidation of glucose-6-phosphate.

85

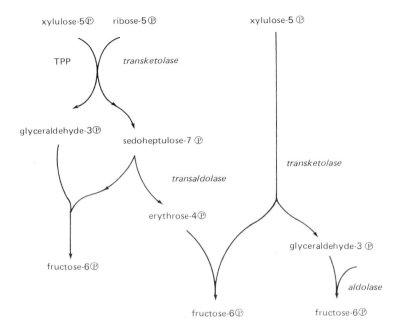

Figure 4.9 The pentose phosphate cycle. The fate of three pentose phosphates only is shown. If three others react in the same way the net result is the conversion of six pentose phosphates to five hexose phosphates. Aldolase catalyses formation of one fructose-6-℗ from *two* glyceraldehyde-3-℗ molecules.

Figure 4.9 shows how six 5-carbon sugars can be converted to five 6-carbon sugars by the combined activity of the enzymes transketolase, transaldolase and aldolase. This means that the overall reaction for six glucose-6-phosphate molecules is:

$$6\ \text{Glc6P} + 12\ \text{NADP}^+ \longrightarrow 5\ \text{Glc6P} + 6\ CO_2 + 12\ \text{NADPH} + 12\ H^+$$

These NADPH can be reoxidised by the respiratory chain providing 36 ATP molecules (or 35 starting from glucose). It is important to consider, however, that the primary role of the pentose phosphate cycle may well be in biosynthesis, not oxidative energy metabolism.

Photosynthesis

Photosynthesis is one of the most important biological processes since it is that in which energy from sunlight is harnessed to the fixation of CO_2 into organic compounds. It occurs in all green plants, algae and in several bacterial genera, autotrophic organisms independent of an external supply of organic compounds for their energy requirement.

Like the aerobic respiration process, there are two sets of reactions. *Light reactions* are concerned with converting light energy into ATP bond energy and

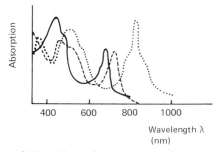

Figure 4.10 Absorption spectra of intact photosynthetic organisms, solid line *Chlorella* (algae), broken line *Chlorobium* (green S bacterium), and dotted line *Chromatium* (purple S bacterium)

producing reducing equivalents as NADPH. The others, the *dark reactions* are those in which CO_2 is reduced to organic compounds; these can occur in the absence of light.

There are major differences between the various groups of photosynthetic organisms. Eukaryotes and the prokaryotic blue-green algae produce both ATP and NADPH in light-dependent reactions and obtain their reducing equivalents from the photolysis of water. Photosynthetic bacteria, however, produce only ATP in light-dependent reactions. They do not evolve oxygen since water is not the electron donor for generation of NADPH, and carry out photosynthesis only in strictly anaerobic conditions. While the blue-green algae resemble eukaryotic algae in their photosynthetic apparatus, in all other aspects of their physiology they have counterparts in bacterial species, and several groups of microorganisms including the Flexibacteria could be considered to be either bacteria or colourless blue-green algae.

The photosynthetic bacteria fall into three major groups: the purple sulphur bacteria (*Thiorhodaceae*), the purple non-sulphur bacteria (*Athiorhodaceae*) and the green sulphur bacteria (*Chlorobacteriaceae*). In the sulphur bacteria H_2S and S^{2-} act as reducing agents:

$$H_2S + NADP^+ \longrightarrow S + NADPH + H^+$$

In some bacteria the elemental sulphur accumulates as granules, in others it is further oxidised to sulphate. The mechanisms of these reactions are not well understood, but they can occur in the absence of light and are probably analogous to those found in the chemolithotrophic organisms discussed later (p. 90). The *Athiorhodaceae* are organotrophic, metabolising organic material such as lactate or malate as electron donors and carbon sources. Not all species of this group are strict anaerobes and obligate phototrophs, the metabolic fate of the substrate differing between aerobic and anaerobic situations.

Differences in light absorption by representative photosynthetic organisms are shown in Fig. 4.10. The purple bacteria absorb at higher wavelength than other groups, a fact which can be used in their isolation by appropriate use of light filters.

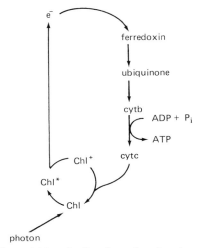

Figure 4.11 Cyclic photophosphorylation, chlorophyll (Chl) is excited by a photon to produce an excited molecule Chl*. This can ionise by transfer of an electron to form the chlorophyll ion (Chl⁺).

Light Reactions

In bacterial systems (excluding the blue-green algae) there is a single photochemically active system which functions in light extending from visible wavelengths to 920 nm in infra-red. The reaction pathway in this case involves a cyclic photophosphorylation of ATP as outlined in Fig. 4.11. The important points are:

i Absorption of a photon of light excites the chlorophyll molecule and an electron is ejected at a high redox potential. In the membranous thylakoids containing the photosynthetic system these electrons are transferred to a reactive site from which they move through a cyclic transport system coupled to the phosphorylation of ADP in much the same way as during oxidative phosphorylation.

ii No NADPH is produced by this cyclic electron flow.

iii Light of higher energy than that absorbed by bacteriochlorophylls can contribute to photosynthesis since there are carotenoids and other accessory pigments which absorb at shorter wavelengths and transmit the energy to the bacteriochlorophylls.

Photosynthesis in blue-green algae and in all eukaryotes is characterised by the presence of two coupled photosystems outlined in Fig. 4.12. In these:

i Light System I is similar to that found in bacteria, but the electron ejected from the chlorophyll (in this case chlorophyll *a*) is used to reduce $NADP^+$; ATP is not formed during this transfer.

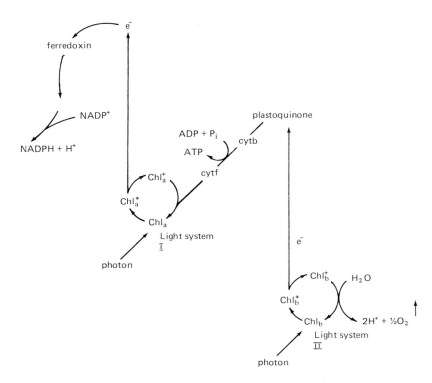

Figure 4.12   Non-cyclic photophosphorylation

ii   The electron from chlorophyll *a* is replaced by Light System II. This system involves higher energy photons than System I, and there is further synthesis of ATP during transfer of this electron from System II to System I.

iii   The systems are coupled by cytochrome *f*, unique to photosynthetic systems. Other components of the chain with the exception of plastoquinone are present in non-photosynthetic organisms.

iv   The chlorophyll of System II is reduced by the splitting of water with the production of oxygen.

v   This non-cyclic system can adjust to differences in the demand for NADPH relative to ATP by uncoupling the two systems, with System I operating in a cyclic way to produce ATP.

Dark Reactions

Fixation of CO_2 occurs initially by addition to ribulose-1,5-diphosphate leading to the formation of two molecules of 3-phosphoglycerate (PGA). This requires the enzyme *carboxydismutase* found in all photosynthetic organisms and chemolithotrophs.

$$CO_2 + H_2O + \begin{array}{c} CH_2O \text{ⓟ} \\ | \\ C=O \\ | \\ -OH \\ | \\ -OH \\ | \\ CH_2O \text{ⓟ} \end{array} \longrightarrow \left[\begin{array}{c} CH_2O \text{ⓟ} \\ | \\ {}^-OOC-C-OH \\ | \\ -OH \\ | \\ -OH \\ | \\ CH_2O \text{ⓟ} \end{array} \right] \prec \begin{array}{c} CH_2O \text{ⓟ} \\ | \\ CHOH \\ | \\ COO^- \\ \\ COO^- \\ | \\ CHOH \\ | \\ CH_2O \text{ⓟ} \end{array}$$

ribulose-1,5-diphosphate   PGA

The PGA so formed is reduced in an ATP-dependent series of reactions analogous to the reversal of the glycolytic pathway:

$$\begin{array}{c} CH_2O \text{ⓟ} \\ | \\ CHOH \\ | \\ COO^- \end{array} \quad \xrightarrow{\text{ATP ADP}} \quad \begin{array}{c} CH_2O \text{ⓟ} \\ | \\ CHOH \\ | \\ COO \text{ⓟ} \end{array} \quad \xrightarrow[P_i]{\text{NADPH NADP}^+ + H^+} \quad \begin{array}{c} CH_2O \text{ⓟ} \\ | \\ CHOH \\ | \\ CH=O \end{array}$$

PGA   1,3-diphosphoglycerate   glyceraldehyde-3-phosphate

It should be noted however that NAD^+ acts as cofactor during glycolysis, and two separate enzymes are present for the two different processes. Thus far one molecule of CO_2 has been fixed with the expenditure of two ATP and two NADPH. From the 3-carbon sugar, a series of sugar interconversions occurs catalysed by the enzymes of the pentose phosphate cycle, *transketolase* and *transaldolase*. These lead to the resynthesis of ribulose-1,5-diphosphate (with expenditure of another ATP) and for every three molecules of CO_2 fixed, six NADPH and nine ATP are expended and one molecule of glyceraldehyde-3-phosphate is formed.

Lithotrophic Energy Production — Oxidation of Inorganic Compounds

Some bacteria, *chemolithotrophs*, can obtain their energy from oxidation of inorganic compounds. In all strains examined, with one exception, the energy release is coupled to ATP biosynthesis via a respiratory chain system. Some of these organisms are autotrophic, fixing CO_2 for their carbon requirement, others are heterotrophic, while some can vary between both modes of life.

An example of this type of organism is *Nitrobacter* which oxidises nitrite to nitrate. The electrons from this oxidation are donated directly to a c-type cytochrome and are transferred to molecular oxygen:

$$\left. \begin{array}{l} NO_2^- + H_2O \\ NO_3^- + 2H^+ \end{array} \right\rangle 2e^- \longrightarrow \text{cyt } c \longrightarrow \text{cyt } a \cdot a_1 \xrightarrow{\text{ADP ATP}} \left\langle \begin{array}{l} H_2O \\ \tfrac{1}{2}O_2 + 2H^+ \end{array} \right.$$

$$E_0' = +0.54 \text{ V}$$

In these organisms the redox potential of the NO_2^-/NO_3^- couple is inadequate to reduce the $NADP^+$ needed for CO_2 fixation. This problem is overcome by an

A

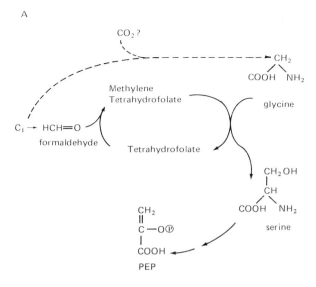

B

Figure 4.13   Assimilatory C_1 metabolism; A, serine
pathway; B, allulose pathway

ATP-dependent reversal of oxidative phosphorylation leading to the reduction of
$NADP^+$. For most inorganic oxidations the oxidation potential is too low for coupling
electron transfer to NAD^+ or flavoproteins, and cytochromes act as acceptors. With
the hydrogen bacteria *Hydrogenomonas* (an aerobic facultative autotroph) the
transfer is however to NAD^+ and a more extensive electron chain operates:

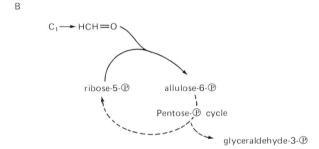

In *Hydrogenomonas* there are two enzyme systems capable of metabolising
hydrogen. A membrane-bound enzyme *hydrogenase* catalyses:

$$H_2 \longrightarrow 2H^+ + 2e^-$$

91

G

Table 4.4   Metabolism of inorganic compounds by chemolithotrophs

Bacteria	Electron donor	Electron acceptor	Metabolic products
Nitrobacter	NO_2^-	Oxygen	NO_3^-, H_2O
Nitrosomonas	NH_4^+	Oxygen	NO_2^-, H_2O
Hydrogenomonas	H_2	Oxygen	H_2O
Desulphovibrio	H_2	Sulphate	H_2O, S^{--}, SO_3^{--}
Micrococcus denitrificans	H_2	Nitrate	NO_2^-, H_2O
Thiobacillus denitrificans	S	Nitrate	SO_4^{--}, N_2
T. thiooxidans	S	Oxygen	SO_4^{--}, H_2O
T. ferrooxidans	Fe^{++}	Oxygen	Fe^{+++}, H_2O

the reducing equivalents are passed on to NAD^+, the other enzyme is a soluble *hydrogen dehydrogenase* catalysing:

$$H_2 + NAD^+ \longrightarrow NADH + H^+$$

The NADH formed by either mechanism is then available to reduce $NADP^+$ for the assimilation of CO_2.

When these bacteria are grown chemolithotrophically, the presence of hydrogen suppresses utilisation of organic substrates such as fructose in a manner analogous to catabolite repression by glucose in other bacteria; enzymes of the Entner-Doudoroff pathway are not synthesised. Mutants lacking hydrogenase, unable to grow chemolithotrophically, do not show this effect and use fructose equally well in hydrogen or air atmospheres; the repression is therefore due to the products of hydrogenase and not hydrogenase *per se*. The hydrogen bacteria are unusual in that they can undergo *mixotrophic* growth in the presence of hydrogen, oxygen and organic substrates: this involves the simultaneous assimilation of CO_2 (p. 89) and organic substrates.

Table 4.4 indicates a few of the types of chemolithotrophic organism found and the nature of the inorganic electron donors and acceptors they employ. In some heterotrophic *Thiobacillus* species, ATP is probably generated in a substrate-level phosphorylation. This involves the formation of the sulphur-containing nucleotide adenosine-5′-phosphosulphate (APS) from sulphite and AMP, catalysed by an Fe^{3+}-containing flavoprotein:

$$2\ SO_3^{2-} + 2\ AMP \xrightarrow{\text{APS reductase}} 2\ APS + 4e^-$$

$$2\ APS + 2\ P_i \xrightarrow{\text{ADP sulphurylase}} 2\ ADP + 2\ SO_4^{2-}$$

$$2\ ADP \xrightarrow{\text{adenylate kinase}} ATP + AMP$$

One-Carbon Compounds

One group of organisms, all prokaryotes, uses organic carbon compounds as energy source, but shows marked similarities to lithotrophs. They metabolise 1-carbon

Table 4.5 Microbial utilisation of one-carbon compounds

Compound	Used by	Growth conditions	Energy source
CH_4	*Methylococcus* etc.	Aerobic	$CH_4 + 2O_2 \longrightarrow CO_2 + 2H_2O$
CH_3OH	*Pseudomonas, Vibrio, Hyphomicrobium*	Aerobic	$2CH_3OH + 3O_2 \longrightarrow 2CO_2 + 4H_2O$
	Methanosarcina	Anaerobic	$4CH_3OH \longrightarrow 3CH_4 + CO_2 + 2H_2O$
CH_3NH_2	*Pseudomonas, Vibrio*	Aerobic	$2CH_3NH_2 + 3O_2 \longrightarrow 2CO_2 + 2NH_3 + 2H_2O$
$HCOOH$	*Pseudomonas* etc.	Aerobic	$2HCOOH + O_2 \longrightarrow 2CO_2 + 2H_2O$
	Rhodopseudomonas palustris	Anaerobic	Photosynthesis
	Methanobacterium	Anaerobic	$4HCOOH \longrightarrow 3CO_2 + CH_4 + 2H_2O$
CO	*Hydrogenomonas*	Aerobic	$2CO + O_2 \longrightarrow 2CO_2$
CO_2	*Hydrogenomonas*	Aerobic	$2H_2 + O_2 \longrightarrow 2H_2O$
	Beggiatoa		$H_2S + 2O_2 \longrightarrow H_2SO_4$
	Thiobacillus		$H_2S_2O_3 + H_2O + 2O_2 \longrightarrow 2H_2SO_4$
	Ferrobacillus		$4Fe^{++} + 4H^+ + O_2 \longrightarrow 4Fe^{+++} + 2H_2O$
	Chlorobium		Photosynthesis
	Purple sulphur and non-sulphur bacteria	Anaerobic	Photosynthesis
	Thiobacillus denitrificans		$5S + 6NO_3^- + 2H_2O \longrightarrow 5SO_4^{--} + 3N_2 + 4H^+$

compounds such as methane, methanol, methylamine and formate as their main energy source and also assimilate these compounds into organic compounds required for biosynthesis (Table 4.5). Some of these organisms will not grow on glucose or amino acid mixtures, others are facultative.

Most 1-carbon compounds are oxidised to CO_2, the reducing equivalents from some oxidation steps being transferred to respiratory chains to provide ATP. For example, in methane oxidisers the pathway appears to involve the following steps:

$$CH_4 \xrightarrow[\substack{O_2}]{oxygenase} CH_3OH \longrightarrow HCH=O \xrightarrow[\substack{NAD^+ \quad NADH \\ + H^+}]{}$$

$$HCOOH \xrightarrow[\substack{NAD^+ \quad NADH \\ + H^+}]{} CO_2$$

Associated with this ability is the capacity to synthesise from the intermediates of these oxidations a 3-carbon compound which can function in the normal pathways of cellular biosynthesis (assimilation of C compounds is discussed in more detail in Chapter 5). If the 1-carbon compound were oxidised to CO_2, then re-reduced as in other lithotrophs, much more energy would be consumed.

There are two alternative pathways of carbon assimilation in 1-carbon utilisers, outlined in Fig. 4.13.

The serine pathway The key step in this is the hydroxymethylation of glycine to serine, involving methylene-tetrahydrofolate generated from one of the 1-carbon intermediates, probably formaldehyde. The involvement of 1-carbon-folate derivatives has been confirmed by the inhibitory effect of sulphanilamide and other folate antagonists (see Chapter 2). The mechanism whereby glycine is regenerated is not fully understood.

The allulose pathway This has an affinity with the ribulose diphosphate cycle of CO_2 incorporation found in other lithotrophs, the important step is the condensation of formaldehyde and ribose-5-phosphate to give allulose-6-phosphate which is then handled via the pentose phosphate cycle.

ENERGY STORAGE

In the presence of excess nutrients microorganisms usually divert some of their metabolism to synthesise compounds which can be broken down during periods of starvation to release energy. These reserves supply the *'maintenance energy'* required by non-proliferating cells to maintain viability, or for spore formation in sporulating organisms. These reserves are usually not tapped until intracellular pools of amino acids and nucleotides can no longer be maintained.

Large amounts of reserve carbon compounds are also formed when nitrogen sources or other essential ions are limiting, but carbon and energy substrates are still available.

Carbohydrate Reserves

Trehalose, a non-reducing disaccharide, is found as a storage compound in fungi and some blue-green algae. In yeasts and slime moulds it is present with glycogen, and is accumulated during vegetative phases of growth. During development leading to spore formation these reserves are used to supply energy. *Glycogen* is found in a large number of microorganisms, and considerable amounts accumulate, especially when there is a nutritional imbalance in energy-rich conditions. The mechanism of glycogen synthesis differs in prokaryotes from that in eukaryotes; in prokaryotes ADP-glucose is the glucosyl donor rather than UDP-glucose (see p. 130).

Lipid Reserves

Poly-β-hydroxybutyrate (PHB) is the most common lipid storage compound in prokaryotes, accumulating in large amounts in photosynthetic bacteria, blue-green algae and many other species. In the larger *Bacillus* species, including the *B. megaterium* and *B. cereus* groups, PHB is the main storage compound supporting sporulation. PHB is not found in eukaryotes, where neutral lipids accumulate. PHB is a linear polymer; its breakdown requires proteolytic activation of the granules followed by hydrolysis by PHB depolymerase to a dimer, subsequently split to the monomer.

Polyphosphates

Polymetaphosphates are found in *volutin* granules in many microorganisms, both eukaryotic and prokaryotic. This polymer acts as both a phosphate and an energy reserve since the phosphoanhydride bond has a high free energy of hydrolysis. Volutin is formed from ATP, and undergoes breakdown in the presence of ADP.

Sulphur

Some sulphur bacteria such as *Thiothrix* and *Beggiatoa* store sulphur produced by the oxidation of H_2S. When sulphide in the medium is exhausted the intracellular sulphur granules are further oxidised to sulphate, supplying the reducing equivalents for photosynthesis or oxidative phosphorylation.

5 Monomer synthesis

In the previous chapter we examined the various pathways by which carbon substrates were oxidised to produce energy and, in some cases, reduced pyridine nucleotides. These pathways are also important since they provide intermediates at a variety of oxidation levels for the biosynthesis of all other molecules in the cell. For this reason they are often referred to as *central pathways.*

Many of the remaining biosynthetic pathways are directed towards forming the precursors of the three main groups of polymers: nucleic acids, proteins and polysaccharides, and also lipids. These precursors include approximately five purines and pyrimidines, twenty amino acids, twenty carbohydrates and about twenty fatty acids. In addition there are pathways for other essential molecules required in lesser amounts, such as pyridine nucleotides, flavins, quinones, porphyrins, coenzyme A, thiamine pyrophosphate, biotin and other cofactors.

Microorganisms vary in their ability to synthesise these low molecular weight precursors and cofactors. Some, including autotrophic species and many commonly used laboratory organisms, can synthesise the total requirement from a simple medium containing one carbon substrate and inorganic sources of N, P, S and ions needed in smaller amounts. At the other extreme are those with a very restricted biosynthetic capacity; *Leuconostoc mesenteroides*, for example, requires virtually all the amino acids, purines, pyrimidines and cofactors. Where microbes are capable of synthesising these compounds they usually use the same sequence as other organisms, and in only a few cases has there been evolution of distinct pathways. Many pathways, particularly those for amino acids, are branched, with early intermediates common to the synthesis of several metabolites. This poses interesting questions as to how these pathways are controlled, and these are discussed in Chapter 7.

Methods for Investigating Biosynthetic Pathways

The earliest biochemical pathways elucidated were those concerned with carbohydrate breakdown and energy metabolism. These were gradually worked out by the chemical identification of intermediates, often using metabolic inhibitors to cause their accumulation. As each intermediate was identified, enzymes were isolated which could act on it, and eventually the overall sequence was determined.

This is now more easily done due to the availability of *isotopic* methods for labelling, which show which compounds act as precursors to others, and from

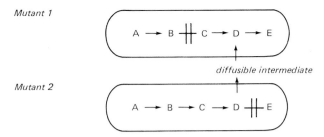

Figure 5.1   Determining the reaction sequence of a metabolic pathway by mutant crossfeeding. Mutant 2 can crossfeed mutant 1, but mutant 1 cannot crossfeed mutant 2.

pulse-labelling studies with rapid sampling after labelling, some details of the reaction sequence can be discerned. Moreover, *mutation* provides a convenient way of specifically affecting each of the enzymes of a reaction pathway. *Auxotrophic* mutants (those unable to synthesise a particular essential metabolite) can be isolated provided the end product of the pathway can be supplied exogenously. These mutants can be defective in any one of the enzymic steps unique to the biosynthesis of the particular metabolite. In some of these, intermediates accumulate and may be excreted into the medium. By 'crossfeeding' tests it is possible to order the sequence of intermediates in the biosynthetic pathway since it is only possible to crossfeed an earlier-blocked mutant with a later-blocked one, as indicated in Fig. 5.1.

REPLENISHMENT (ANAPLEROTIC) PATHWAYS

A number of the central metabolites involved in energy metabolism are not synthesised in substrate amounts by the energy-yielding pathways. The most important examples are the TCA cycle intermediates. Several of these are however, also starting points in the biosynthesis of nearly half of the amino acids and pyrimidines. Several mechanisms therefore exist for conversion of other central intermediates formed in substrate amounts to TCA cycle intermediates, these depend on the organism and the nature of the carbon substrate used for growth.

In aerobic organisms growing on *acetate* the main mechanism is the *glyoxylate cycle*; this is found in many bacteria, fungi, algae and protozoa and is summarised in Fig. 5.2. It is characterised by the presence of two enzymes: *isocitrate lyase*, catalysing the cleavage of isocitrate to succinate and glyoxylate, and *malate synthase* which catalyses the condensation of acetyl-CoA and glyoxylate to form malate. The net result of these reactions taken with the TCA cycle is the conversion of two acetyl-CoA to one dicarboxylic acid of the TCA cycle.

For organisms growing on 3-carbon compounds (pyruvate, lactate) or substrates leading to these (sugars, glycerol) another system operates, involving the fixation of CO_2. In most microorganisms there are a number of enzymes for fixing CO_2 to either pyruvate or phosphoenolpyruvate with the synthesis of either malate or

97

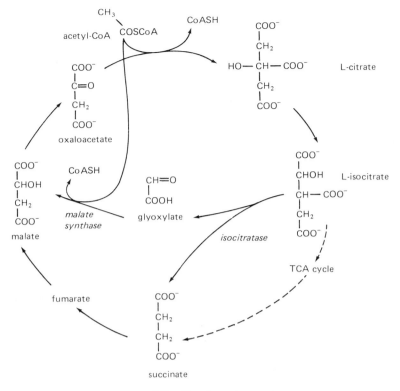

Figure 5.2   The glyoxylate cycle

oxaloacetate. In yeast and *Arthrobacter globiformis* the anaplerotic role appears to be fulfilled by the enzyme *pyruvate carboxylase* catalysing:

$$
\begin{array}{c}
CH_3 \\
| \\
C=O \\
| \\
COO^-
\end{array}
\quad
\xrightarrow[\substack{biotin \\ CO_2}]{\substack{ATP \quad ADP + P_i}}
\quad
\begin{array}{c}
COO^- \\
| \\
CH_2 \\
| \\
C=O \\
| \\
COO^-
\end{array}
\quad + H^+
$$

pyruvate   oxaloacetate

In enterobacteria, however, studies with mutants have shown another enzyme, *phosphoenolpyruvate carboxylase* operates in this role

$$
\begin{array}{c}
CH_2 \\
\| \\
C-O-\textcircled{P} \\
| \\
COO^-
\end{array}
\quad
\xrightarrow[\substack{P_i}]{\substack{CO_2}}
\quad
\begin{array}{c}
COO^- \\
| \\
CH_2 \\
| \\
C=O \\
| \\
COO^-
\end{array}
\quad + H^+
$$

98

ASSIMILATION OF NITROGEN AND SULPHUR

Many organic metabolites contain N, S and P. These are generally obtained from inorganic sources, and their assimilation to an organic form is an essential step in the biosynthesis of many monomers. Phosphorus is usually taken up as phosphate in the medium and converted into organic phosphate in familiar reactions.

Nitrogen can be assimilated as organic N, or from inorganic sources such as NH_4^+, NO_2^-, NO_3^- or, for a few very specialised bacteria and including blue-green algae, as molecular N_2. NH_4^+ is usually directly incorporated in a reductive addition to α-oxoglutarate to give glutamate. A second NH_4^+ can be used in the conversion of glutamate to glutamine:

| α-oxoglutarate | glutamate | glutamine |

The amine group of glutamate can be transferred directly to other α-ketoacids (e.g. pyruvate) to form other amino acids by transamination.

Nitrate is reduced by some organisms not as a terminal electron acceptor in energy metabolism, but to NH_4^+ for biosynthesis. This is carried out in a series of reductive steps for which the first step only has been characterised. This is catalysed by *nitrate reductase,* the reducing equivalents being transferred from NADPH:

$$NO_3^- \xrightarrow[\text{NADPH} \quad \text{NADP}^+ + \text{H}^+]{\text{fp; Mo}} NO_2^- \longrightarrow ? \longrightarrow [NH_2OH] \longrightarrow NH_4^+$$

Two types of microorganism fix atmospheric N_2: the genus *Rhizobium* found in close symbiotic association with leguminous plants and normally only fixing N_2 in the root nodules of legumes; and free-living blue-green algae and bacteria such as the aerobic species *Azotobacter,* the facultative anaerobe *Klebsiella aerogenes* (which only fixes under strictly anaerobic conditions) and the strict anaerobe *Clostridium pasteurianum.* No eukaryotic microorganism capable of fixing N_2 is known.

Energy required for N_2 fixation is derived chemotrophically from fermentation or respiration, or in the case of the blue-green algae, phototrophically. The process is highly endergonic, needing an estimated fifteen moles of ATP to fix each mole of N_2, most of this being used to generate hydrogen and a suitable reductant. Special low electrode potential electron carriers have been identified in these microorganisms, including the most widespread, *ferredoxin,* a non-haem iron protein. In *Azotobacter* ferredoxin can be replaced by a *flavoprotein,* while in

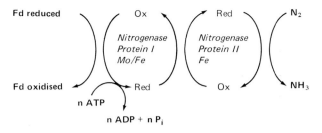

Fd = Ferredoxin

Figure 5.3   The microbial fixation of nitrogen

Clostridium pasteurianum flavodoxin can function in cells deprived of iron and unable to form ferredoxin. The net process is:

$$N_2 + 3XH_2 \longrightarrow 2NH_3 + 3X,$$

where X is the low potential electron carrier. The enzyme system catalysing N_2 fixation is composed of two proteins of which one contains both Mo and Fe, the second has Fe only. This system is relatively non-specific and can reduce a number of substrates containing triple bonds, including acetylene, cyanide, nitrous oxide and azide, e.g.

$$HC \equiv CH \longrightarrow H_2C{=}CH_2$$

$$HC \equiv N \longrightarrow CH_4 + NH_3$$

The system is unusual in that acetylene, azide and cyanide are usually very toxic to biological systems due to their high affinity for transition metal ions. The scheme for N_2 fixation is shown in Fig. 5.3. In media containing sources of fixed nitrogen, the nitrogenase system is repressed — the molecular basis of this repression by ammonia is not yet fully understood. In aerobic N_2-fixers some compartmentation of these enzymes probably exists since highly reduced conditions are required, and the fixation is strongly inhibited *in vitro* by oxygen.

Sulphur is usually taken up as sulphate. Studies of cysteine-requiring mutants of various bacterial and fungal species have shown that sulphate is first activated by adenylylation to form the nucleotide adenylyl sulphate (APS) (Fig. 5.4), then phosphorylated and reduced to sulphite and sulphide in a series of steps:

$$SO_4^{2-} \xrightarrow{\text{ATP} \quad P_i} APS \xrightarrow{\text{ATP} \quad ADP} 3'\text{-PAPS} \xrightarrow{\qquad PAP \qquad} $$

$$\text{NADPH} + H^+ \quad \text{NADP}^+$$

PAPS: phosphoadenylylsulphate
PAP: 3'phosphoAMP

$$SO_3^{2-} \xrightarrow{\qquad\qquad} S^{2-}$$

$$\text{NADPH} + H^+ \quad \text{NADP}^+$$

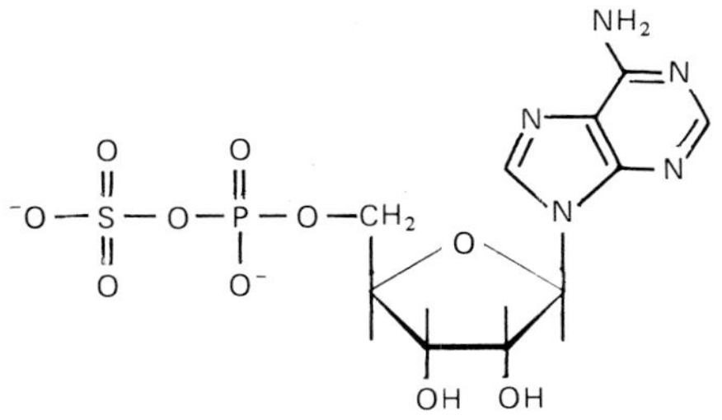

Figure 5.4   Adenylyl sulphate (APS)

In yeasts and other fungal systems the final reaction is the condensation of serine and sulphide to form cysteine:

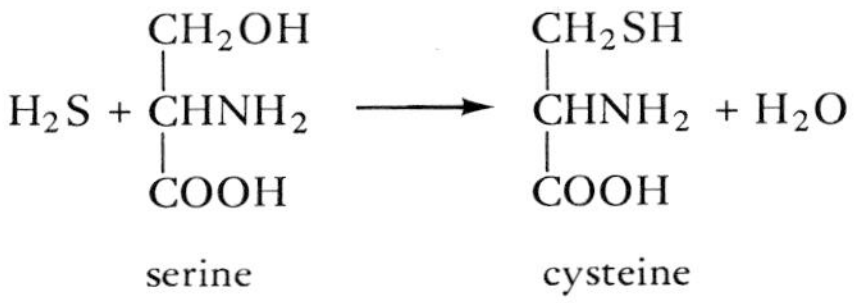

$$H_2S + \begin{array}{c} CH_2OH \\ | \\ CHNH_2 \\ | \\ COOH \end{array} \longrightarrow \begin{array}{c} CH_2SH \\ | \\ CHNH_2 \\ | \\ COOH \end{array} + H_2O$$

      serine   cysteine

In the *Enterobacteriaceae*, a different reaction occurs, involving O-acetylation of serine and subsequent reduction with sulphide:

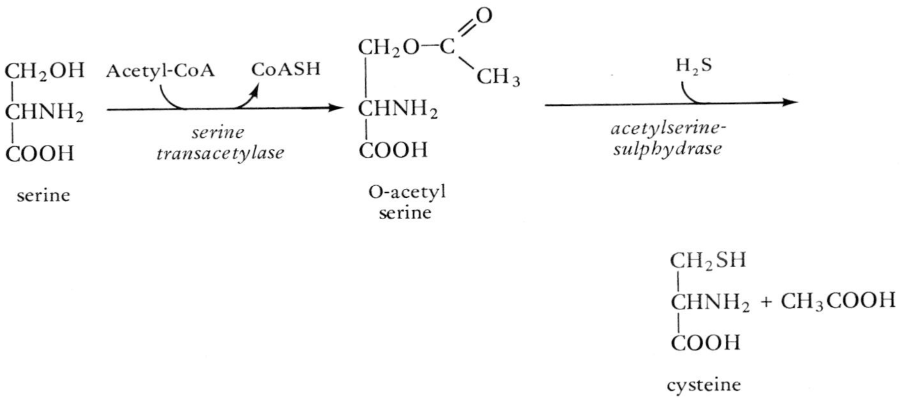

$$\begin{array}{c} CH_2SH \\ | \\ CHNH_2 \\ | \\ COOH \end{array} + CH_3COOH$$

      cysteine

BIOSYNTHETIC PATHWAYS

It is not the purpose of this book to describe in detail the biochemical pathways leading to synthesis of all low molecular weight metabolites. In most cases these reactions are common to most organisms and extensive descriptions are available in general biochemistry texts. Here we are interested in general patterns of the flow of central metabolites to various important compounds, and the synthesis of some metabolites which are unique to microorganisms. A general scheme for the position of central metabolism in biosynthesis is outlined in Fig. 5.5.

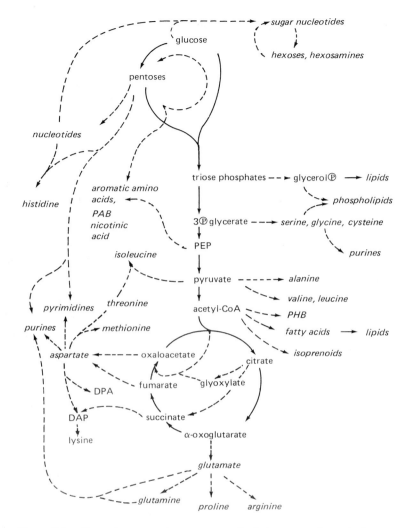

Figure 5.5   General pattern of monomer synthesis from central metabolites

Nucleotide Synthesis

Nucleotide precursors of nucleic acids are composed of a purine or pyrimidine base glycosidically linked to ribose or deoxyribose phosphates:

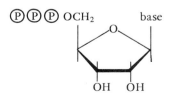

The ribose phosphate moiety is derived from the pentose phosphate cycle, and the deoxypentose is not formed until a later stage in nucleotide synthesis. A common step in synthesis of both purines and pyrimidines is the activation of ribose-5-phosphate by phosphorylation to phosphoribosyl pyrophosphate (PRPP):

Pyrimidines   PRPP condenses with a base precursor, *orotic acid*, which is formed in a series of reactions commencing with the condensation of carbamoyl phosphate and aspartate. The enzyme catalysing this step, *aspartate transcarbamylase*, has been extensively studied since it forms an important site of feedback control of pyrimidine nucleotide synthesis. Orotic acid is formed by a cyclisation reaction; it condenses with PRPP leading to UMP, and thence UDP and UTP which can be enzymically converted to CTP.

Purines   The synthesis of these differs from the pyrimidines in that the bases are built up completely on the ribose-5-phosphate moiety by a lengthy series of steps involving the amination of PRPP and subsequent additions of glycine, amino groups, C_1 groups and aspartate to give inosine monophosphate. From IMP divergent pathways exist for forming AMP and GMP.

The nucleotide triphosphates are all formed by successive phosphorylations from ATP, catalysed by *nucleotide phosphotransferase*. The deoxyribose of deoxyribonucleotides is formed at the nucleotide diphosphate stage by an NADPH-linked reduction:

Deoxythymidine phosphates are formed from deoxyCMP via deoxyUMP as an intermediate:

These above reactions are of interest in that the free bases do not participate in the normal synthetic reactions. In some organisms there are pathways by which added adenine, thymine or uracil can be taken up and incorporated into nucleic acid. This is fortunate since it enables the specific labelling of RNA and DNA in many bacteria. In some eukaryotic microorganisms, including yeasts, one of the enzymes *thymidine kinase* needed for incorporating thymine or thymidine is lacking, and since dTMP is taken up poorly, the specific labelling of DNA is a problem.

Amino Acid Synthesis

From Fig. 5.5 it can be seen that most of the amino acids are formed from a few main central metabolites. They are therefore often grouped into families on the basis of their precursors (Table 5.1). Histidine, the sole exception, is synthesised from PRPP and ATP in a sequence of reactions in which the heterocyclic ring is derived from the adenine of ATP.

Lysine Biosynthesis   L-lysine is an essential amino acid in the human diet, and therefore methods have been developed to produce it industrially by fermentation. This amino acid also provides an example of separate evolution of two completely different pathways for synthesis of the same compound. In fungi,

Table 5.1   Families of amino acid synthesis

Family	Precursors	Amino acids
aromatic amino acids	erythrose-4-phosphate and phosphoenolpyruvate	→ tryptophan → tyrosine → phenylalanine
pyruvate	pyruvate	→ alanine → valine → leucine
glutamate	α-ketoglutarate	→ glutamine → glutamate → proline → arginine → lysine (fungi)
aspartate	oxaloacetate	→ lysine (bacteria) aspartate → threonine → isoleucine → methionine
serine	3-phosphoglycerate	→ serine → cysteine → glycine

Histidine is the sole amino acid member of its pathway, but its precursors, PRPP and ATP, are involved in nucleotide synthesis.

104

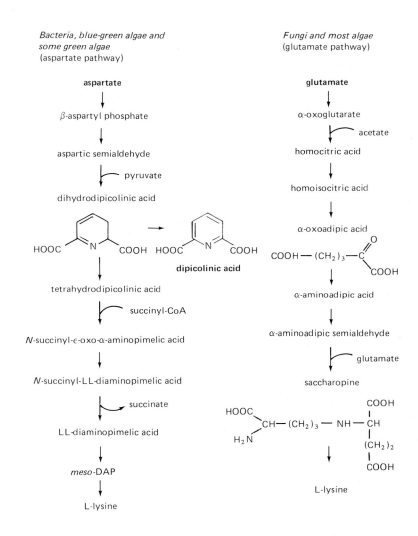

Figure 5.6 Alternate pathways of lysine biosynthesis

including Phycomycetes, Ascomycetes and Basidiomycetes, and in a few green algae, including *Euglena,* L-lysine is formed from L-glutamate via α-aminoadipate. In bacteria, blue-green and most green algae another important pathway exists starting from *aspartate* and proceeding *via* diaminopimelic acid (DAP). The two sequences are shown in Fig. 5.6.

The aspartate pathway for lysine is essential to prokaryotes since it provides two compounds unique to their physiology: *diaminopimelic acid* occurring in some peptidoglycans, and *dipicolinic acid* formed by the oxidation of dihydrodipicolinic acid, a major constituent of bacterial spores.

105

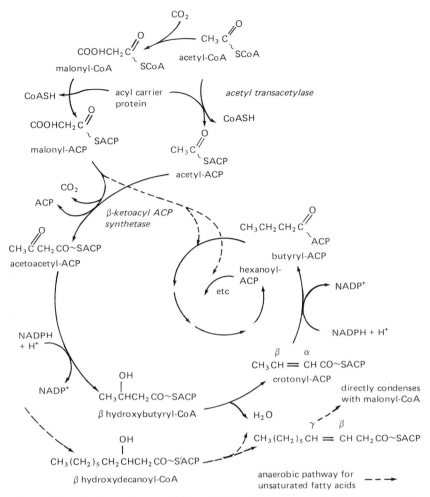

Figure 5.7 The biosynthesis of fatty acids. The important reaction in the anaerobic pathway of unsaturated fatty acid synthesis is indicated

Fatty Acid Synthesis Various fatty acids are found in microbial cells as components of lipids, but unlike the precursors of proteins and nucleic acids, they do not occur as free acids. Fatty acids may be unsaturated or saturated, and relatively large amounts of the former are found in psychrophiles where their lower melting point is essential in maintaining the fluidity of the cell lipids.

The key compounds in fatty acid synthesis, and hence in lipid formation, are derivatives of coenzyme A and *acyl carrier protein* (ACP). ACP has now been isolated from numerous microorganisms since its initial discovery in *Clostridium kluyveri*, an anaerobe capable of growth on a mixture of acetate and ethanol as carbon source. The protein contains β-alanine and 2-mercaptoethylamine and is bound to

106

the fatty acyl moiety at all stages of the biosynthesis as indicated in Fig. 5.7. In prokaryotes the ACP and associated enzymes are present in soluble form, whereas in eukaryotes they form tightly bound particulate *fatty acid synthetase* complexes. In all microorganisms the mode of saturated fatty acid synthesis is the same, involving the addition of 2-carbon fragments in the form of acetyl groups, and their subsequent reduction. The initiating acetyl group is carboxylated to malonyl-CoA in a biotin-dependent reaction; thereafter, two enzymes convert acetyl-CoA and malonyl-CoA to their respective ACP derivatives. The higher fatty acids are then built up by the cyclic action of four enzymes as indicated in Fig. 5.7.

As well as saturated fatty acids, most microbial lipids contain a range of unsaturated fatty acids, including monounsaturated ones such as oleic acid $[CH_3-(CH_2)_7-CH=CH-(CH_2)_7COOH]$ and those containing two or three double bonds. Two mechanisms are known for the biosynthesis of unsaturated fatty acids. The *aerobic pathway*, involves molecular oxygen, NADPH and the acyl-CoA derivative of the appropriate acid. The double bond is always inserted between C_9 and C_{10} regardless of the length of the fatty acid chain:

$$CH_3-(CH_2)_{14}-CO-SCoA \xrightarrow[\frac{1}{2}O_2 \quad H_2O]{} CH_3-(CH_2)_5-\overset{10}{CH}=\overset{9}{CH}-(CH_2)_7-CO-SCoA$$

palmitoyl-CoA   palmitoleyl-CoA

This process occurs in eukaryotes and in a number of aerobic bacteria, including *Bacillus megaterium* and *Micrococcus lysodeikticus.* The other pathway, in which there is no requirement for oxygen, is found in anaerobic bacteria, and a number of aerobes, for example *Pseudomonas, Lactobacillus* and blue-green algae. This *anaerobic pathway* is a branch of normal saturated fatty acid synthesis introduced at the β-hydroxydecanoyl-ACP level (at the 10-carbon fatty acid stage). Instead of the normal α,β dehydration, a β,γ dehydration occurs. Longer chain unsaturated fatty acids are then formed by further addition of 2-carbon fragments. This means that the products of the anaerobic pathway differ from those of the aerobic in the position of the double bond: e.g. the C_{18} products are *cis*-vaccenic acid (Δ 11) and oleic acid (Δ 9) respectively.

Branched chain fatty acids occur in many organisms. Their synthesis can be initiated from corresponding branched chain amino acids such as leucine or isoleucine:

D(+)-12-methyl
tetradecanoate

Alternatively, a methyl group can be added to unsaturated fatty acids:

$$CH_3(CH_2)_7CH=CH(CH_2)_7COOH \qquad \text{oleic acid}$$

$$CH_3(CH_2)_7\underset{\underset{CH_2}{\|}}{C}CH_2(CH_2)_7COOH \qquad \text{10-methylene stearic acid}$$

$$CH_3(CH_2)_7\underset{\underset{CH_3}{|}}{C}HCH_2(CH_2)_7COOH \qquad \text{10-methyl stearic acid}$$

Isoprenoids

A number of terpene derivatives are also found in microbial cells. These are composed of isoprenoid units having the general structure:

$$-(CH_2-\underset{\underset{CH_3}{|}}{C}=CH-CH_2)-$$

Among the polyisoprenoid compounds identified are the carrier lipids involved in bacterial peptidoglycan and lipopolysaccharide formation (Chapter 6), coenzyme Q and vitamin K, and different carotenoids present in several groups of microorganisms. Those found in photosynthetic bacteria and in many bacteria isolated from soil resemble hydrocarbons found in plant tissues:

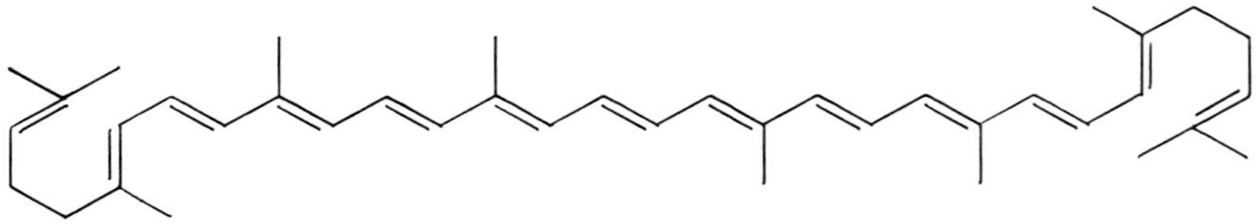

They appear to be absent from anaerobic, non-photosynthetic bacteria. Synthesis of the isoprenoid compounds proceeds from acetyl-CoA by way of acetoacetyl-CoA to mevalonic acid, which is phosphorylated by ADP and decarboxylated to yield *isopentenyl pyrophosphate*. In a further series of reactions, the C_{15} compound, farnesyl pyrophosphate, is formed from three isopentenyl pyrophosphate residues. From farnesyl pyrophosphate, sequential addition of isopentenyl pyrophosphate gives the C_{55} isoprenoid alcohol pyrophosphate involved in the synthesis of several polymers found outside the cell membrane (Chapter 6). In eukaryotes, dolichols — C_{90} isoprenologues — perform the same function. Farnesyl pyrophosphate also undergoes condensation in eukaryotic cells to form squalene and, subsequently, *sterols*, in a reaction which is absent from most prokaryotes, but is probably found in some squalene-containing methane-utilising bacteria. Farnesyl pyrophosphate is also the precursor for parts of the chlorophyll and quinone molecules.

108

Lipid and Phospholipid Synthesis

Lipids are essentially derivatives of glycerol:

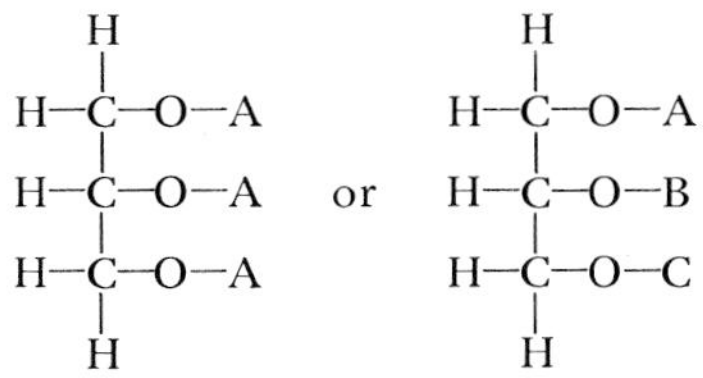

$$
\begin{array}{ccc}
& H & & H \\
& | & & | \\
H-C-O-A & & H-C-O-A \\
& | & & | \\
H-C-O-A \quad or & & H-C-O-B \\
& | & & | \\
H-C-O-A & & H-C-O-C \\
& | & & | \\
& H & & H
\end{array}
$$

In both prokaryotic and eukaryotic cells, phospholipid synthesis requires cytidine triphosphate (CTP). The reaction between CTP and an L-α-phosphatidic acid produces a cytidine diphosphate diglyceride:

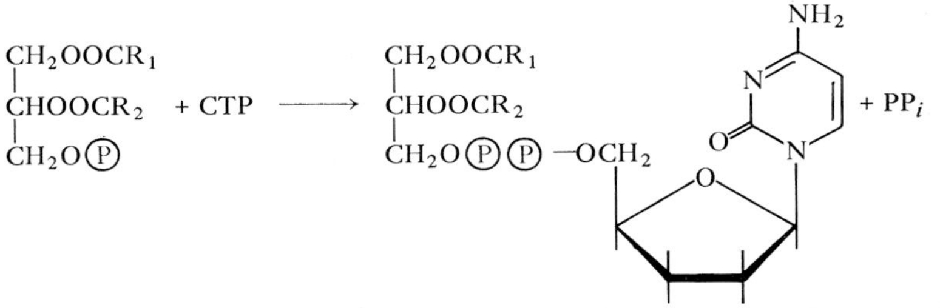

$$
\begin{array}{l}
CH_2OOCR_1 \\
| \\
CHOOCR_2 \quad + CTP \longrightarrow \\
| \\
CH_2O\,\textcircled{P}
\end{array}
\qquad
\begin{array}{l}
CH_2OOCR_1 \\
| \\
CHOOCR_2 \\
| \\
CH_2O\,\textcircled{P}\,\textcircled{P}-OCH_2
\end{array}
\qquad + PP_i
$$

Triglyceride formation is accomplished by the dephosphorylation of L-α-phosphatidic acid by a phosphatase enzyme. The D-α, β-diglyceride then reacts with the acetyl-CoA derivative of the appropriate fatty acid to form a triglyceride.

In bacteria, phosphatidyl ethanolamine is synthesised *via* the intermediate phosphatidyl serine:

CDP-diglyceride + L-serine $\longrightarrow$ phosphatidyl serine + CMP

Phosphatidyl serine $\longrightarrow$ phosphatidyl ethanolamine + CO$_2$

This mechanism differs from that found in eukaryotic cells, where there is direct interaction between CDP-ethanolamine and a diglyceride:

CDP-ethanolamine + D-α,β-diglyceride $\longrightarrow$

phosphatidyl ethanolamine + CMP

Monosaccharides

In microorganisms monosaccharides are seldom, if ever, found as the free sugars. Usually they are present either as components of polysaccharides or other polymers, or in smaller amounts in the cytoplasm as sugar phosphates and 'sugar nucleotides' (nucleoside diphosphate monosaccharides). Monosaccharides and their derivatives can be synthesised from the common sugar substrates: glucose, fructose

109

or mannose. Microorganisms vary widely in their ability to utilise external sugars; some such as *E. coli* can take up such sugars as arabinose, and even the methyl pentose rhamnose, while others show a much more restricted range. Some can even assimilate sugar phosphates or sugar nucleotides without degradation. For those microorganisms growing on substrates such as lipids, amino acids, or other acids, monosaccharides are synthesised by some of the Embden-Meyerhof pathway enzymes acting in reverse, with a shunt around the energetically unfavourable pyruvate → phosphoenolpyruvate step *via* one of the enzymes converting malate or oxaloacetate to phosphoenolpyruvate.

After absorption, monosaccharides can be phosphorylated by the non-specific enzyme *hexokinase* at the expense of ATP. Thus glucose-6-phosphate is formed from glucose and converted to glucose-1-phosphate by *phosphoglucomutase.* In enterobacteria the sugar is usually phosphorylated during transport into the cell via the phosphoenolpyruvate-dependent reaction sequence discussed in Chapter 3. This type of mechanism is absent from those organisms fermenting glucose via the Entner-Doudoroff pathway.

Sugar Nucleotides   Before they can participate in many biosynthetic or interconversion reactions, sugar phosphates are activated by reaction with nucleoside triphosphates to form 'sugar nucleotides'. These are primarily synthesised from hexose phosphates such as α-glucose-, α-galactose- or α-mannose-1-phosphate in the presence of the appropriate *UDP-Glc pyrophosphorylase* and nucleoside triphosphate, for example:

$$\text{glucose-1-phosphate} + \text{UTP} \longrightarrow \text{UDP-glucose} + \text{PP}_i$$

The reaction is reversible, and the pyrophosphorylases are specific for both the hexose phosphate and the nucleoside triphosphate, and thus provide a means of regulating synthesis of several polymers containing the same monosaccharide. A wide variety of 'sugar nucleotides' are found in microbial cells, but the function of some of them remains obscure; most cells contain UDP-glucose, UDP-*N*-acetylglucosamine and GDP-mannose. The role of such compounds in polymer synthesis is understood when one considers that hydrolysis of one mole of the glucose phosphate bond of UDP-glucose yields 7.6 kcalories.

By contrast, hydrolysis of the phosphate bond of α-D-glucose-1-phosphate yields only 4.8 kcalories.

The nucleoside-diphosphate monosaccharides serve two purposes in the microbial cell; they provide a mechanism for interconversion of certain monosaccharides and they are the glycosyl donors needed for polysaccharide formation. The glycosyl moiety of the sugar nucleotides can be transformed by a variety of mechanisms, some of which are confined to prokaryotes (Fig. 5.8). Epimerisation of carbon atom 4 converts UDP-glucose to UDP-galactose in the presence of UDP-galactose-4-epimerase. Correspondingly, UDP-glucuronic acid, UDP-*N*-acetyl glucosamine and UDP-arabinose are epimerised to UDP-galacturonic acid, UDP-*N*-acetylgalactosamine and UDP-xylose respectively. The cofactor in these reactions is NAD$^+$; it is probable that the sugar is first oxidised then reduced at the

110

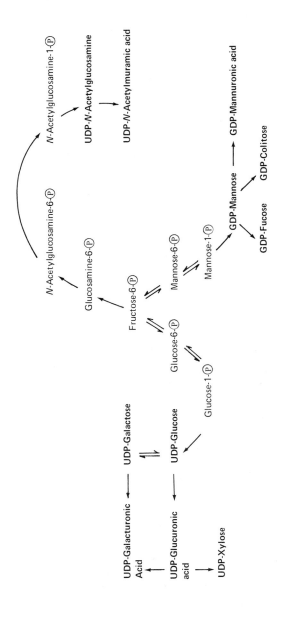

Figure 5.8 Microbial interconversions of sugar nucleotides

carbon 4 position. Uronic acids are formed from the corresponding neutral hexose portion of a 'sugar nucleotide' by oxidation at the carbon 6 position:

$$\text{GDP-mannose} \xrightarrow[\text{NAD}^+ \quad \text{NADH} + \text{H}^+]{\substack{\textit{GDP-mannose} \\ \textit{dehydrogenase}}} \text{GDP-mannuronic acid}$$

The pentose xylose, in the form of UDP-xylose is formed by decarboxylation of UDP-glucuronic acid. Although this reaction is found in a number of bacteria, it is more common in yeasts. This is as yet the only known mechanism of formation of 'activated' pentoses other than pentose phosphates.

One of the reactions which is confined to prokaryotes is that by which UDP-N-acetylglucosamine combines with phosphoenol pyruvate to yield UDP-N-acetylmuramic acid, the precursor of the muramic acid pentapeptides involved in prokaryotic wall biosynthesis (Chapter 6):

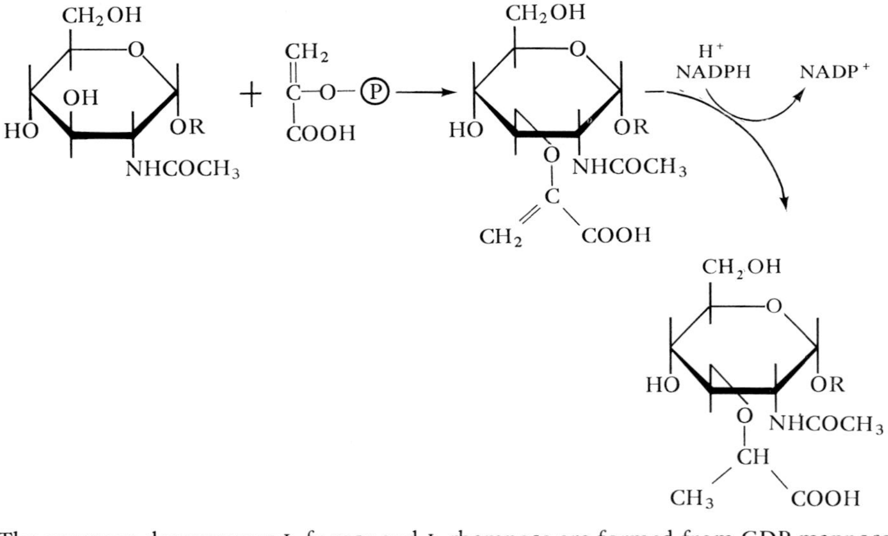

The common deoxysugars L-fucose and L-rhamnose are formed from GDP-mannose and TDP-glucose respectively. The hexose portion of the sugar nucleotide loses a molecule of water to form an intermediate which is oxidised in the presence of NAD^+ to give a nucleoside diphosphate 4-keto-6-deoxyhexose. Further rearrangement leads to the formation of the L-6-deoxyhexoses. Similar mechanisms are involved in forming the 3, 6-dideoxyhexoses — tyvelose, abequose, paratose and colitose — which are found exclusively in the cell wall lipopolysaccharides of Gram-negative bacteria.

Summary

Microorganisms show a very great variation in their nutritional requirements, that is some are very exacting while others can synthesise all their metabolic needs from a minimal medium containing one carbon source and essential salts.

112

Pathways for monomer biosynthesis are very similar from one group of organisms to another, one exception being lysine biosynthesis. Differences have, however, evolved in the way various groups of organisms regulate these pathways, particularly the branched ones, and these are discussed briefly in Chapter 7. The wide diversity of biosyntheses carried out by microorganisms makes generalisation difficult, but the following observations can be made. Most of these pathways require activation of intermediates and therefore the expenditure of energy. With lipid synthesis and monosaccharide formation in particular, the intermediates are activated not only for synthetic reactions, but also for interconversions such as desaturation, isomerisation or oxidation.

6 Polymer synthesis and metabolism

Polymers are required by microbial cells for a wide variety of functions. Some are restricted to a single general function. DNA acts as an information storage molecule, peptidoglycan in bacteria and cellulose, mannans, glucans and chitin in fungi have a structural role in providing mechanical resistance to osmotic lysis, while glycogen, poly-β-hydroxybutyrate and polyphosphate appear to act as storage or regulatory polymers which maintain substrates in a form inaccessible to degradative enzymes and exerting negligible osmotic pressure. Other polymers have been adapted in the course of evolution to a variety of functions. Certain proteins, such as those in ribosomes, may be structural in function, while most are enzymes with specific catalytic activity. Others function in transport of molecules across cell membranes (permeases). Still others provide cells with the means of locomotion. Diversity of function is illustrated by RNA, which plays intermediary (mRNA), adaptor (tRNA) and scaffolding (rRNA) roles in protein synthesis.

Structure of Polymers

In chemical terms microbial polymers can be classified as *homopolymers* – composed of a single subunit; and *heteropolymers*, composed of a number of different subunits. Few homopolymers are found in microorganisms; they include bacterial glycogen, levans and dextrans, poly-β-hydroxybutyrate and poly-D-glutamic acid, polyphosphate and glucan and mannan of fungal cell walls. Heteropolymers are more common; they include many polysaccharides, bacterial peptidoglycan and all proteins and nucleic acids. The sequence of monomers in heteropolymers can be of two types. There may be a repeating group of subunits as in most polysaccharides and peptidoglycan. For extracellular polysaccharides these vary from disaccharides to hexasaccharides, while the backbone chain of peptidoglycan is a repeating unit of N-acetylmuramyl-N-acetylglucosamine. Repeating units of small size may be essential in polymers formed external to the microbial cell membrane. The alternative type of polymer, found in proteins and nucleic acids, has an ordered structure without repeating units. In these the monomer order is determined by a genetic template.

For many polymers the three-dimensional arrangement of subunit monomers is important. Thus peptidoglycan consists of a polysaccharide backbone crosslinked by short peptide chains effectively forming an extended network of high strength which determines the shape of the bacterial cell. In bacterial spores this crosslinking

114

is considerably reduced, leaving free many ionised groups on the peptide which may play an important part in maintaining the resistant, dormant structure.

The molecular architecture of proteins is fundamental to their activity. The primary sequence of amino acids adopts a secondary structure of helical and non-helical regions. These are then organised into a tertiary structure in which ionic, hydrophobic and covalent bonds all play some part. For some enzymes this tertiary structure is sufficient to provide the correct molecular environment for catalysis, in others a quaternary structure composed of polypeptide subunits is necessary. In most cases the tertiary structure of proteins is that most favoured thermodynamically and is adopted spontaneously. For other polymers synthesis of the three-dimensional form depends on a pre-existing framework. A good example is peptidoglycan which is synthesised by additions to terminal groups of available peptidoglycan; complete removal of the cell wall peptidoglycan by lysozyme prevents regeneration of new cell wall material by the protoplasts formed.

General Characteristics of Polymer Metabolism

While the diversity of polymers implies variety in mechanisms for their synthesis there are a number of general characteristics common to most:

i   Macromolecular synthesis is a highly endergonic process. In most cases energy is derived from hydrolysis of high energy bonds; the substrates either contain these (as in ATP) or are activated prior to the polymerisation step.
ii   Requirement for a mechanism *initiating* polymerisation. In some cases this involves a primer molecule.
iii   Requirement for *termination* processes.
iv   A *template* is needed for those heteropolymers with ordered, non-repeating subunits to specify the correct monomer arrangement.

Another common characteristic of many microbial polymers is their susceptibility to rapid breakdown as well as synthesis. For example, growth requires controlled lysis of wall components to enable insertion of new growing points. A molecule in a dynamic state due to synthesis and degradation acting simultaneously is undergoing *turnover*. In general, synthesis and degradation are distinct processes with different enzymes responsible. Furthermore, associated with this ease of breakdown of polymers there must be either 'compartmentation' to separate them from potential degradative enzymes, or these enzymes must be regulated such that they act only in certain conditions.

NUCLEIC ACIDS

Nucleic acids are linear heteropolymers synthesised from four main nucleotide subunits (in DNA: dATP, dCTP, dGTP and dTTP; in RNA: ATP, CTP, GTP and UTP),

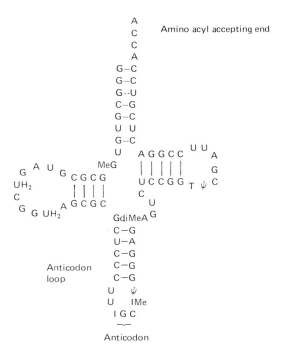

Figure 6.1 Yeast alanine transfer RNA showing intrachain base pairing and secondary structure

linked by phosphodiester bonds between the 3′ and 5′ positions of the ribose or 2-deoxyribose residues. Other modified nucleotides do occur in minor amounts; ribosomal RNA (rRNA) contains methylated nucleotides, minor bases are found in transfer RNA (tRNA) and certain bacteriophages have substituted deoxyribonucleotides in their DNA (glucosylated in T4), protecting the phage DNA from breakdown by lytic (restriction) enzymes synthesised by the phage or host.

Everyone is by now familiar with the pairing by hydrogen bonding of purine and pyrimidine bases (A=T, G≡C), the double helix of DNA, and the genetic implications of this. It is often not appreciated that many RNA molecules also show a high degree of secondary structure maintained by base pairing. This is best understood in tRNAs (Fig. 6.1) in which the rigid arrangement of the anticodon loop provides the anticodon triplet with the appropriate energy of binding to the

116

codon in messenger RNA (mRNA) during protein synthesis. This secondary structure may occur in mRNA to act as a control element (Chapter 8).

DNA

The metabolic fates of DNA are best understood in terms of its function as the information storage tape coding for the total genetic potential of the cell. When cells divide, the total genetic information must also duplicate and separate in synchrony. This is DNA replication. Another process of considerable importance is DNA repair. Errors and chemical damage can arise in DNA, particularly under extreme environmental conditions. These defects can lead to loss of cell viability unless efficiently repaired. A third fate of DNA also involves enzymic modification of the molecule. During genetic recombination a process of strand exchange occurs. Recombination is relatively infrequent in prokaryotes (although involvement of a recombination type of process in repair is common); in eukaryotes it is an essential feature of the meiotic phase of their life cycle.

Replication

Major differences exist between prokaryotes and eukaryotes in the arrangement and replication of DNA. In those few bacterial species examined in sufficient detail one circular chromosome of circumference 1 to 2 mm is found although several smaller circular DNA plasmids may be present. In eukaryotic miroorganisms the nuclear DNA is organised into chromosomes (four in haploid *Hansenula*, seven in *Neurospora* and seventeen in haploid *Saccharomyces* species). Each chromosome has a centromere region which is the point of attachment for DNA to the spindle apparatus during its segregation in mitosis or meiosis.

DNA replication is not a simple process; in *E. coli* at least eight genes are needed. Replication begins at a single region on the chromosome, the origin, and proceeds bidirectionally from there (see Chapter 2). At least one protein and probably two are involved in this initiation step, and one at least must be synthesised for each cell division cycle. The resultant replication intermediate, a three-armed closed loop structure with two loops of equal length, has been visualised by autoradiography and is shown in Fig. 6.2. DNA segregation into daughter cells, after replication has taken place, is probably mediated by chromosome attachment to the bacterial cell membrane and it seems that both the origin and the replication point(s) are attached to the membrane or mesosomes.

Polymerisation during replication involves a rather more complicated sequence of reactions than was at first expected:

i   The two helical strands serving as templates must be separated. From prokaryotes and at least one eukaryote (smut fungus; *Ustilago maydis*) an 'unwinding protein', which might serve this function, has been isolated.

ii   All known DNA polymerases catalysing the reaction:

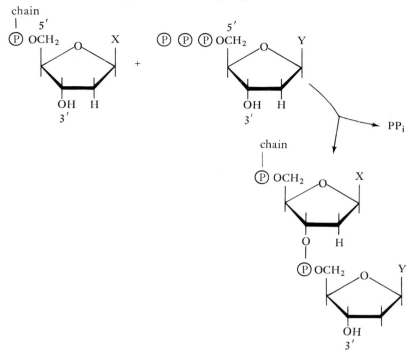

synthesis is in the $5' \rightarrow 3'$ direction only (travelling on a template from its $3'$ end to its $5'$ end). DNA is however antiparallel, having one strand with $5' \rightarrow 3'$ polarity the other with $3' \rightarrow 5'$, so that at the replication fork only one of the two complementary strands can be read in the same direction as replication is proceeding. On the other hand it is known from autoradiography that recently replicated DNA does not have long single strand regions. The problem is overcome by DNA polymerase travelling in fairly short steps on the antiparallel strand away from the fork as shown in Fig. 6.3. This generates short pieces of newly synthesised DNA.

iii   A third enzyme DNA ligase, is needed to join these fragments into two continuous complementary strands.

iv   Replication DNA polymerase (III) requires a primer molecule on which to catalyse nucleotide condensation. This need not be DNA, but for some phage at least can be RNA. This therefore implicates a number of other enzyme activities (not necessarily on separate enzyme molecules) including possibly an *RNA polymerase*, a *nuclease* capable of excising primer RNA from an RNA/DNA hybrid, and another DNA *polymerase* (I) to fill the gaps.

The possible overall replication process is summarised in Fig. 6.3.

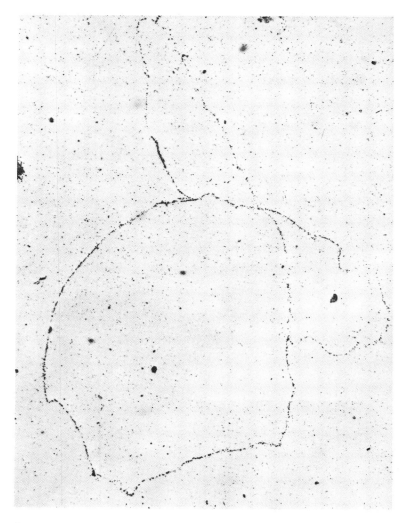

Figure 6.2 Autoradiograph of the replicating *E. coli* chromosome from a cell labelled with [^{3}H]-thymidine for two generations. (Courtesy of Dr. J. Cairns and the Cold Spring Harbor Laboratory)

Eukaryotic Replication

In bacteria replication proceeds at a rate of about 1000 monomers per second. Eukaryotic microorganisms on the other hand replicate at about one tenth this rate, and contain at least five to ten times as much DNA as *E. coli*. Despite this, under optimal conditions they can duplicate their DNA in roughly the same time as *E. coli*. They accomplish this by having multiple replication origins on each chromosome such that during DNA synthesis there are many replication forks at

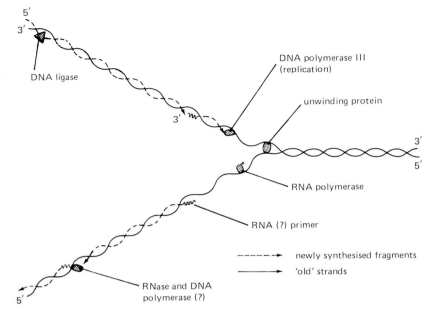

Figure 6.3 Diagrammatic representation of bacterial DNA replication

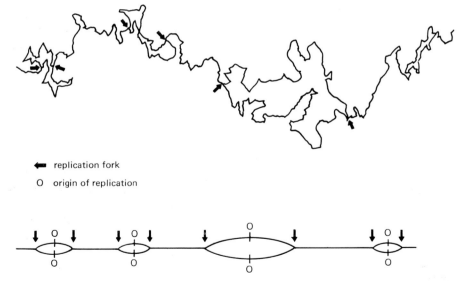

Figure 6.4 Yeast DNA in the process of replication. Direct tracing of an electron micrograph, and a diagrammatic representation are shown. (Tracing from Peters, T. D., Newton C. S., Byers, B., & Fangman, W. L. (1973) *Cold Spring Harbor Symp. Quant. Biol.* **38**, 9.)

any one time. This has been visualised by autoradiography in higher eukaryotes and by direct electron microscopic study in yeast (Fig. 6.4). Replication from each origin is bidirectional, with adjacent origins located between 15 and 70 μm apart. In yeast this corresponds to about 5 to 20 origins on an average-sized chromosome.

It should be emphasised that we have dealt with but two of the possible modes of DNA replication. Other models have been proposed to account for replication in certain bacteriophages and eukaryotic organelles known to contain DNA.

Finally, some viruses containing RNA as their genetic material are capable of coding for an enzyme synthesising DNA on this RNA template. This presumably enables integration of viral information into the animal host replication machinery in much the same way as temperate phage form lysogens (p. 160) with their bacterial host. This RNA-dependent DNA polymerase, called reverse transcriptase, is an important enzyme since these viruses may be implicated in cancer.

DNA Repair and Recombination

In most organisms the replication DNA polymerase is not the only enzyme with DNA-synthesising activity; in *E. coli* two other enzymes have been isolated. These other enzymes participate in a range of repair and recombination processes in conjunction with a large number of other enzymes modifying DNA. In yeast there are at least six pathways for DNA repair, each with its own set of activities. This is not surprising when one considers the range of DNA damage which is repaired: single or double-strand breaks, UV-induced thymine-thymine dimer formation, base mismatching and strand crosslinking induced by chemical mutagens. An idea of the range of activities modifying DNA during repair can be gained from Fig. 6.5.

Recombination has been studied mainly by geneticists, and at least thirteen models have been proposed to account for the phenomenon. It is clear however that it occurs mainly through breakage and reunion of DNA strands, and thus nucleases and ligases are involved. Some repair functions are also common to recombination since single mutations affecting both processes have been found.

Degradation

From a close look at Fig. 6.5 it is obvious that some enzymes degrading DNA have specificities associated with their function. Some are endonucleases introducing 'nicks' into a single strand; others are exonucleases removing nucleotides sequentially from one end. One repair enzyme in *E. coli*, DNA polymerase I, has two distinct activities. One is a $5' \rightarrow 3'$ exonuclease acting at a 'nick' to remove one strand of the DNA duplex; the other a DNA polymerase resynthesising DNA against the remaining strand. This obviously represents an efficient method of error removal and may also function in primer excision during DNA replication.

Certain bacteria, and in some cases phage themselves, produce enzymes for degrading bacteriophage DNA. Some of these *restriction enzymes* are highly specific endonucleases which cleave DNA at sites specified by a short sequence of bases. The purified enzymes find considerable application in study of DNA

121

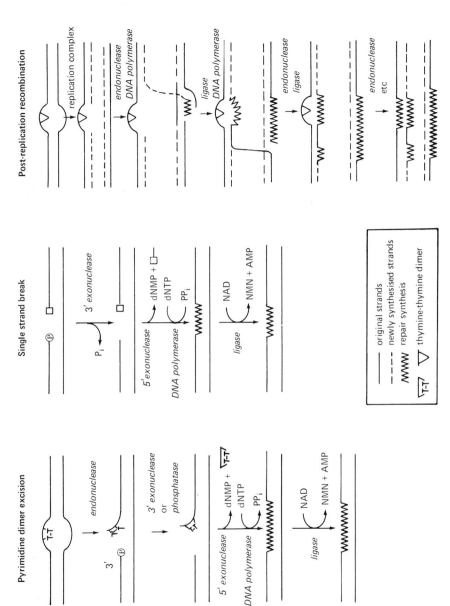

Figure 6.5 Proposed mechanisms for DNA repair, illustrating the involvement of enzymes capable of specifically modifying DNA

sequences, and are proving, in conjunction with ligases and other DNA modifying enzymes, to be essential tools for genetic engineering.

RNA

Synthesis

The three major classes of RNA found in microorganisms differ in size, but all participate in protein synthesis and all are synthesised in a similar way from a DNA template in the process called *transcription*. RNA molecules are essentially single-stranded, but probably all contain some helicity formed by intrachain base pairing.

In prokaryotes only one DNA-dependent RNA polymerase has been isolated, while in yeast and many other eukaryotic microorganisms three or more enzymes can be found. The need for three in eukaryotes is unclear, but is probably related to regulation. One seems to be specific to rRNA synthesis, one functions in transcription of mitochondrial DNA and the other is presumed to form mRNA and tRNA. It is clear that rRNA synthesis is controlled in a different way from the other classes of nuclear RNA. In bacteria there is also differential control, but this is exerted by various proteins which can complement the one enzyme thereby modifying its activity. These polymerases all require the four ribonucleotide triphosphates ATP, CTP, GTP and UTP, double-stranded DNA as template and either Mg^{2+} or Mn^{2+}. The reaction can be represented as:

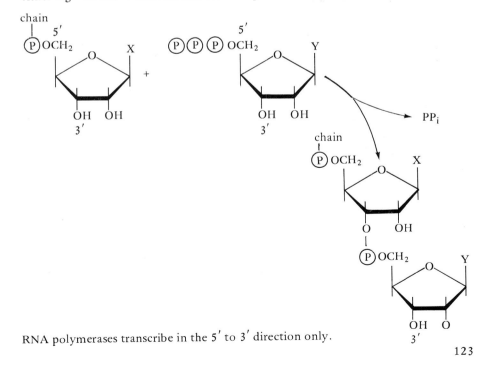

RNA polymerases transcribe in the 5′ to 3′ direction only.

I

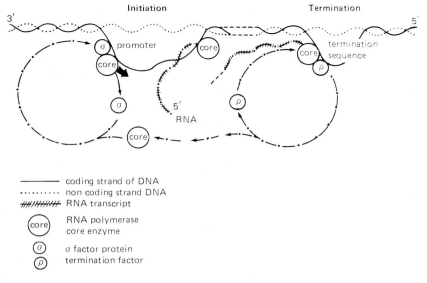

——————— coding strand of DNA
············ non coding strand DNA
##!//##/# RNA transcript
(core) RNA polymerase
core enzyme
(σ) σ factor protein
(ρ) termination factor

Figure 6.6 Bacterial transcription

In vivo RNA synthesis is a highly directed process. The transcripts, whether mRNA, rRNA or tRNA, must begin at a given point on one strand only and finish after the correct sequence has been read. This requires initiation and termination mechanisms. Initiation is so far understood only in bacterial systems. RNA polymerases contain four (*Bacillus subtilis*) or five (*E. coli*) subunit polypeptides which together form the 'core' enzyme. *In vitro*, this enzyme begins transcription randomly from either strand of the DNA template. A further protein, the σ factor, can complement core enzyme, which then begins transcription only from a specific sequence on one strand (the coding strand) of the DNA. This site is a short sequence of bases termed the *promoter*. The σ factor is then released from this complex and can complement a further molecule of core enzyme.

The core enzyme polymerises nucleotides at the rate of about 100 per minute until the appropriate termination sequence is reached. The mechanism of termination is not fully understood but in phage λ is analogous to the reverse of initiation in that another protein, ρ, interacts with the enzyme and the termination sequence on the DNA causing release of the core enzyme. These interactions are summarised in Fig. 6.6.

Post-transcriptional Modification

There are differences between the various RNAs in more detailed aspects of their completion. In bacteria the 70 *S* ribosome contains three pieces of RNA designated 23 *S*, 16 *S* and 5 *S* from their sedimentation behaviour. In eukaryotes these are 28 *S*, 18 *S* and 5 *S*. The 28 *S* and 18 *S* species are not synthesised separately; rather a single long molecule of 45 *S* RNA is produced, then specifically and successively cleaved by endonucleases to produce the mature subunits and excess material which is degraded. During this *processing* some nucleotides are methylated enzymically in either the base or ribose group; this may play some part in the accurate assembly of the ribosome. In bacteria the rRNA cistrons are also arranged in compound transcriptional units in 16 S, 23 S, 5 S array, and the two larger rRNAs may be transcribed as a single 30 S piece. To provide an adequate supply of rRNA to the growing cell, bacteria have at least six sets of rRNA cistrons. In eukaryotic microorganisms there is a much greater degree of gene amplification — yeast has 150 copies. In both systems the transcriptional units are arranged as adjacent sequences on the DNA with short spacers between. This arrangement may be the result of linear amplification of a single gene during evolution together with selective advantages of such an arrangement maintaining them in this state. Closely linked genes are less frequently involved in recombination and there is less likelihood of mutation in one rRNA cistron affecting others. In eukaryotes the rDNA is organised into a specific structure the nucleolus; this is made possible by the close linkage of these genes.

For tRNAs, other enzymes are required for complete synthesis since the three terminal nucleotides (.pCpCpA) at the amino acyl-accepting end are added sequentially without involvement of RNA polymerase. tRNAs also contain bases not usually found in other RNA; these are formed by enzymic modification after transcription.

A number of enzymes have been found which catalyse polyribonucleotide formation in the absence of template DNA. Bacterial polynucleotide phosphorylase catalyses the reaction:

$$nXPP \longrightarrow (XP)_n + nP_i$$

thereby differing from RNA polymerase in using nucleoside diphosphates as precursors, and in producing a random polymer. This enzyme acting in reverse may be involved in phosphorolysis of RNA. In eukaryotes another enzyme of this type produces polyadenylate (poly (A)) sequences. These are attached post-transcriptionally to messenger-like RNA synthesised in the nucleus. The poly (A)-RNA complex then finds its way to the cytoplasm. Poly (A) may therefore be important in transport of mRNA or its protection during transport from its site of synthesis to translation on cytoplasmic ribosomes.

Degradation

RNA degrading enzymes are also known. Some are located within the cell and have quite specific functions. One of the more important of these to unicellular organisms is the endonuclease responsible for rapid breakdown of mRNA. This turnover of mRNA enables the cell to adapt very rapidly to changes in the environment (see Chapter 8). Another in this class is the endonuclease involved in rRNA processing. Other intracellular enzymes break down large amounts of cellular RNA in cells facing starvation, providing a lot of the substrates for endogenous metabolism necessary to keep the cell viable.

Other degradative enzymes are extracellular, either periplasmic (between cell wall and membrane) or are excreted into the medium; these presumably act as scavenger enzymes degrading RNA into units which can cross the cell membrane. In many organisms these are produced mainly under conditions of stress.

PROTEIN

Synthesis

Proteins are linear heteropolymers formed from up to twenty precursors, the amino acids. The order of amino acids is coded in a particular region of DNA, loosely termed a gene, more rigorously a cistron.

Each amino acid is coded by a linear triplet of bases in DNA, three being the minimum needed to code for twenty amino acids. The code is redundant, since there are sixty-four possible triplet combinations thus most amino acids are coded by more than one triplet.

DNA is not read directly but through mRNA, which is transcribed on DNA and acts as the template for protein synthesis. There are many reasons why the existence of such an intermediate is vital to the cell; for example:

i   Each gene can be transcribed many times and hence amplification is achieved.

ii   Control of protein synthesis can be more simply achieved (see Chapter 8).

iii   Messenger breakdown can also play a part in control.

iv   In eukaryotes the existence of an intermediate is probably necessary since DNA is contained within the nucleus while the translation machinery is situated on the endoplasmic reticulum.

As for synthesis of most polymers, activated precursors are needed. Each amino acid is activated in a two step enzymic process requiring ATP and a tRNA molecule specific to that amino acid (Fig. 6.7), tRNAs serve a multifunctional role during translation and are not merely vehicles for accepting amino acids in a reactive state. Amino acids have no affinity for nucleic acids and cannot recognise a codon. The tRNA serves this function, each having on one loop (anticodon loop) of its clover-leaf structure a three base sequence complementary to the coding triplet, see Fig. 6.2 for the yeast tRNA^{Ala}. Some fifty different tRNAs are presumed to exist (one for each codon specifying an amino acid), and for each amino acid there is a separate enzyme charging tRNA.

Amino acyltRNAs do not condense to form peptides in free solution; this takes place on ribosomes. These particles are composed of two subunits, each containing rRNA and twenty to thirty proteins. Bacterial 70 S ribosomes can be dissociated to 50 S (containing 23 S and 5 S RNA, about 30 proteins) and 30 S (16 S RNA and about 20 proteins) subunits. In eukaryotes the ribosome is larger (80 S dissociating to 60 S and 40 S) and is inhibited by different antibiotics to those affecting bacterial ribosomes.

At the ribosome, charged tRNAs are successively brought into conjunction with each other and the mRNA template. Each mRNA is usually being read by many ribosomes at once and polyribosomes containing ten or more ribosomes can be

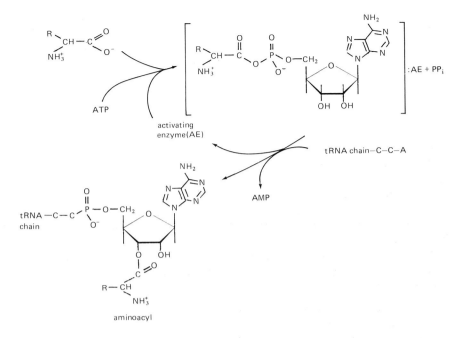

Figure 6.7   Activation of amino acids for protein synthesis

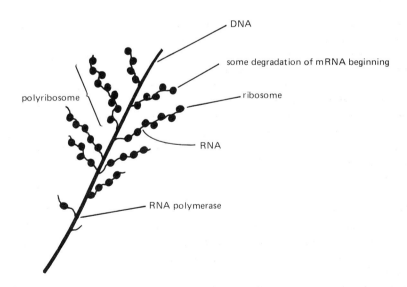

Figure 6.8   Bacterial gene transcription and simultaneous translation on the DNA template seen by electron microscopy of gently lysed bacteria

128

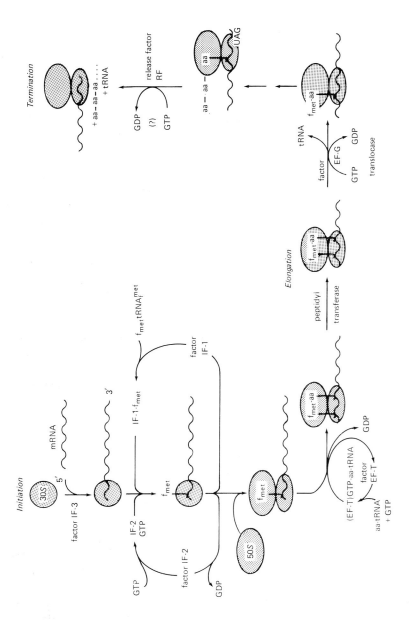

Figure 6.9   Events involved in protein synthesis

isolated from very gently broken cells. In bacteria mRNA is translated as it is synthesised on the DNA template; this has been visualised directly by electron microscopy (Fig. 6.8). Thus translation proceeds from the $5'$ end of the mRNA, corresponding to protein synthesis from the NH_2^- to the COOH— terminal end.

Initiation, Elongation and Termination

As with transcription, translation requires a mechanism for correct initiation and termination. Protein factors also participate in a complicated sequence of events depicted in Fig. 6.9. Initiation occurs at specific codons, AUG and GUG, as long as they are preceded by a ribosome binding sequence on the mRNA of about 20 nucleotides. A four to seven base sequence in this region is complementary to the $5'$ end of the 16 S ribosomal RNA. The initiation codons are recognised by a specific tRNA. In bacteria there are two methionyl tRNAs; one of these, $tRNA_f^{met}$, is charged with methionine. This methionyl $tRNA_f^{met}$ is formylated and is incorporated into the NH_2-terminal position during initiation. In eukaryotes, initiation follows a generally similar pattern, although there is no requirement for N-formylation of the initial methionyl $tRNA_f^{met}$.

Several features of the scheme in Fig. 6.9 should be noted.

i   Ribosomes do not begin as the 70 S form, but are built up from the subunits after the 30 S unit has formed a complex with the template and N-formyl-methionyltRNA$_f^{met}$.

ii   The initiation factor IF-3 is involved in attachment of the mRNA to the 30 S subunit. This factor may select the messages to be read and is therefore a potential site for control of gene expression.

iii   In addition to the three initiation factors there are elongation factors EF-T and EF-G involved with addition of aminoacyl tRNA to the ribosome and with translocation of the message along the ribosome.

iv   The high requirement for GTP during protein synthesis is concerned with the transfer of molecules; ATP is involved in activation of amino acids.

Termination is also directed by triplets; UAA, UAG and UGA have been found by genetic and biochemical studies to act in this way. These triplets can be formed by mutation of others coding for amino acids (nonsense mutations) and in such mutants the polypeptide chain is terminated prematurely.

Protein Modification and Degradation

During synthesis, or just after, the nascent polypeptide folds into its functional configuration. In most cases this seems to occur spontaneously, the polypeptide adopting the thermodynamically most favoured state. In other cases, such as in assembly of the head of phage T4 participation of non-structural proteins is required. One change which occurs is the specific cleavage of precursor poly-peptides by intracellular proteases. At the simplest level, the N-formylmethionyl moiety is often removed. On the other hand, during bacterial sporulation a number of proteins are cleaved by endoproteases thereby gaining an altered enzyme activity or resistance to heat.

Protein turnover also occurs. Microorganisms are capable of rapid adaptation to a change in culture conditions, and cells facing starvation begin to break down some of their protein complement to provide amino acids for synthesis of new enzymes capable of dealing with any alternative substrates which may exist. In bacteria this turnover is not extensive until conditions of nutrient deprivation arise.

POLYSACCHARIDE SYNTHESIS

So far the discussion has centred on synthesis on templates, an essential requirement for gene expression. Many cellular polymers are composed of repeating units for which no direct template is required. However, the ordering of subunits within these chains is due to the specificity of the transferase enzymes adding the subunits in sequence.

There are many polymers which fall into this category; some such as poly-β-hydroxybutyrate have already been discussed (Chapter 5), and most of the essential principles emerge from studying polysaccharides.

Homopolysaccharides

Microbial cells form several types of homopolysaccharide, some being linear molecules while others are branched. The most common is glycogen, the only intracellular polysaccharide found in bacteria. Its synthesis in prokaryotes follows a similar pattern to that in eukaryotes with one major difference, in the nature of the glycosyl donor. In eukaryotes the polymer is formed by sequential addition of glucose units from UDP-glucose; in prokaryotes the enzyme catalysing this reaction shows a primary specificity for ADP-glucose. *In vitro* the activity of this enzyme with ADP-glucose is 100-fold that with UDP-glucose as substrate while other glucose-containing sugar nucleotides are inactive.

Thus the first specific stage in the synthesis of bacterial glycogen is an activation, forming ADP-glucose (see p. 111). Subsequently a linear polymer is formed by *glycogen synthetase* (ADP-glucose: 1,4 glucan-4-glucosyl transferase). A third enzyme, the '*branching enzyme*', transfers a segment of 6 to 8 glucose residues from the main chain to form an α 1 → 6 linkage at a branch point. Mutants lacking either of the first two enzymes fail to form glycogen, while lack of the branching enzyme leads to accumulation of an unbranched polyglucose staining blue with iodine. *Glycogen phosphorylase*, an enzyme long thought to be involved in glycogen synthesis, functions solely in the breakdown of the polymer.

This requirement for ADP-glucose is interesting since it may provide the prokaryotic cell with a way of controlling glycogen synthesis independent of cell wall formation which requires UDP-linked sugars as substrates. No other ADP-linked sugars are known to be of metabolic importance in bacteria.

Much less is known about the synthesis of other homopolysaccharides such as the poly-N-acetylneuraminic acid 'Vi' antigen of *Salmonella* strains or the extracellular cellulose fibres secreted by *Acetobacter* species. These also require sugar nucleotides, but levans and dextrans (polyfructose and polyglucose respectively) are formed by a different mechanism.

130

A specific substrate, sucrose, is required and, in the presence of *levansucrase* or *dextransucrase*, both extracellular enzymes, the polymers are formed:

$$x \text{ Sucrose} + (\text{fructose})_n \longrightarrow (\text{fructose})_{n+x} + x \cdot \text{glucose}$$

$$\quad\quad\quad\quad\quad\quad \text{levan} \quad\quad\quad\quad\quad\quad \text{levan}$$

$$x \text{ Sucrose} + (\text{glucose})_n \longrightarrow (\text{glucose})_{n+x} + x \cdot \text{fructose}$$

$$\quad\quad\quad\quad\quad\quad \text{dextran} \quad\quad\quad\quad\quad\quad \text{dextran}$$

Heteropolysaccharides

Microorganisms synthesise a large number of different types of heteropoly-saccharide, too many to enumerate here. Rather it is better to concentrate on one system: the biosynthesis of the lipopolysaccharide found in cell walls of *Salmonella typhimurium*, which has been most intensively studied and illustrates many features common to heteropolysaccharide synthesis. In particular it provides a clear example of how extracellular polymers are synthesised outside the cell membrane from intracellular components, and typifies the two types of synthetic process in which a 'core' portion of the molecule is formed followed by addition of side chains (the O-antigen determinants in this case).

In formation of the 'core', monosaccharides are transferred sequentially from sugar nucleotide donors (the activated precursors) to an acceptor molecule containing the 8-carbon sugar acid, 2-keto-3-deoxyoctonic acid (KDO), and two moles of heptose. To this acceptor, glucose is transferred from UDP-glucose. In this bacterial species, one or two molecules of galactose are then transferred from UDP-galactose, then a further molecule of glucose and one of *N*-acetyl-D-glucosamine. This completes formation of the core (Fig. 1.9). Mutation of any one of these enzymes or of a sugar nucleotide-synthesising enzyme results in formation of an incomplete polymer; those sugars distal to the block are not added, nor are the side chains. As a result, phage receptors and serological determinants of the cell surface are altered but the viability of the cell is unaffected under laboratory conditions.

'Side-chains' are synthesised on a different receptor molecule which has an important function in all synthesis of extracellular and cell wall polysaccharides. This is a polyisoprenoid lipid phosphate which has been termed *antigen carrier lipid* (ACL):

$$\begin{array}{c} CH_3 \\ \quad \diagdown \\ \quad\quad C{=}CHCH_2(CH_2 \overset{\overset{\displaystyle CH_3}{|}}{C}{=}CHCH_2)_9 CH_2 \overset{\overset{\displaystyle CH_3}{|}}{C}{=}CHCH_2OH \\ \quad \diagup \\ CH_3 \end{array}$$

The primary reaction involving this receptor requires the addition of a *sugar phosphate* from the appropriate sugar nucleotide. Other sugars are added successively by specific transferase enzymes until a repeating unit of the side-chain, containing 3 to 4 sugar residues, is formed. Transfer of repeating units leads to formation of a polymer of increased chain length, polymerisation

131

taking place in such a way that the monosaccharide adjacent to the lipid phosphate is added to the terminal sugar on another oligosaccharide-ACL complex. Thus, in bacteria where the side chain repeating units are trisaccharides of galactose, rhamnose and mannose, the sequence of reactions is:

$$UDP-Gal + ACL-P \longrightarrow ACL-P-P-Gal + UMP$$

$$ACL-P-P-Gal + TDP-Rha \longrightarrow ACL-P-P-Gal-Rha + TDP$$

$$ACL-P-P-Gal-Rha + GDP-Man \longrightarrow ACL-P-P-Gal-Rha-Man + GDP$$

$$ACL-P-P-Gal-Rha-Man + ACL-P-P-Gal-Rha-Man \longrightarrow$$
$$ACL-P-P-Gal-Rha-Man-Gal-Rha-Man + ACL-P-P$$

Build up of the repeating units continues until the normal 'side-chain' of about eight units is formed. Then, as in the case illustrated of a *Salmonella* lipopolysaccharide, the entire side chain is transferred to the lipopolysaccharide 'core' along with release of the lipid pyrophosphate:

$$
\begin{array}{ccccc}
 & & \text{Glc.NAc} & \text{Gal} & \\
 & & | & | & \\
ACL-P-P(Gal-Rha-Man)_n + & Glc & \text{------} & Gal-Glc-Hept \ldots \\
\downarrow & & & & \\
 & & \text{Glc.NAc} & \text{Gal} & \\
 & & | & | & \\
ACL-P-P + (Man-Rha-Gal)_n-Glc & & \text{------} & Gal-Glc-Hept \ldots
\end{array}
$$

The ACL molecules are located in the cell membrane and have the important role of transferring repeating units formed on the internal side of the cell membrane across the membrane to the outside where they are linked into the side chain. Other microbial polysaccharides are also synthesised by processes involving assembly of their repeating unit structures on a similar lipid intermediate. This mechanism appears to be used in synthesis of all such polymers found external to the bacterial cell membrane. Thus the extracellular capsules of the genus *Klebsiella* are formed using a lipid diphosphate; others, such as the mannan of *Micrococcus lysodeikticus* and teichoic acid of *Bacillus licheniformis* are assembled on the lipid phosphate. In fungi cell wall polysaccharides seem to be synthesised in part in intracellular vesicles which coalesce with the membrane at the growing point; a lipid phosphate intermediate has however been identified in yeast and *Aspergillus.*

Peptidoglycan

This characteristic structural polymer of prokaryotic cell walls, is much more complicated than other polysaccharides and indeed has some features of a polysaccharide and others of peptide. Part of the molecule is a repeating carbohydrate unit structure to which is attached peptide subunits of regular sequence (Fig. 1.7). These linear strands are usually crosslinked by a second type of peptide creating a giant network polymer of considerable strength. As might be expected from both its structure and its extracellular location, formation of the

132

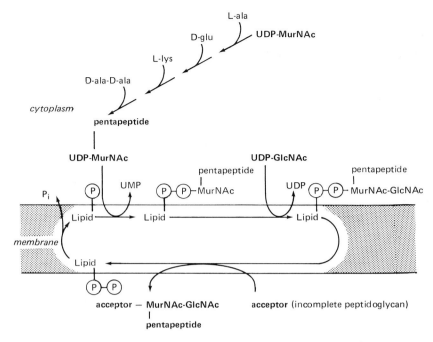

Figure 6.10   Lipid intermediates in peptidoglycan synthesis

peptidoglycan is a complex process. In intact cells the synthesis proceeds in four stages illustrated in Fig. 6.10.

i   Formation of UDP-sugars. These are UDP-*N*-acetyl-D-glucosamine and UDP-*N*-acetylmuramic acid (see Chapter 5).

ii   Amino acids to form the peptides are added sequentially, or in one case as a dipeptide, to the UDP-*N*-acetylmuramic acid to produce a nucleotide sugar pentapeptide.

iii   UDP-*N*-acetylmuramyl pentapeptide is transferred to the isoprenoid carrier lipid and *N*-acetylglucosamine added. This forms the repeating unit of the polymer, which is transported across the membrane and transferred to an acceptor of partially completed polymer. Chains containing from ten to thirty repeating units are formed.

iv   The linear chains are crosslinked. In some Gram-positive bacteria crosslinks are formed directly as an amide bond between the DAP or lysine residue and the penultimate D-alanine of another peptidoglycan chain. In other bacteria cross-linking occurs by addition of a pentaglycine bridge to the penultimate D-alanine of the peptidoglycan precursor with release of the terminal D-alanine. Formation of the peptide bridge involves a tRNA which may be specific to peptidoglycan synthesis, but none of the peptide linkages are formed on templates, their order being determined by the specificity of enzymes needed at each successive step.

133

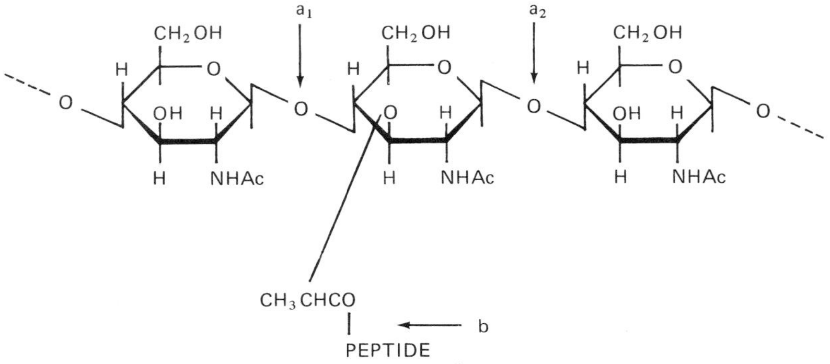

$a_{1,2}$ = Endoglycosidases; a_1 : muramidase; a_2 : N-acetylglucosaminidase
b = Amidase

Figure 6.11   Peptidoglycan as an enzyme substrate

An interesting feature of both peptidoglycan and lipopolysaccharide synthesis is that the isoprenoid lipid phosphate is activated by further phosphorylation. Both a pyrophosphatase and a kinase have been found in *Staphylococcus aureus*, and these may function to control availability of carrier lipid, and possibly the location of cell wall synthesis.

Degradation of Peptidoglycan (Fig. 6.11)

Most, if not all, bacteria produce one or more enzymes capable of depolymerising peptidoglycan; their function in the growth of the cell wall and in cell division has been discussed (p. 37). These can be *amidases* cleaving the peptide chains from the polysaccharide backbone, or *glycosidases* acting at one or other of the bonds between the amino sugars. The best known example of the latter type of enzyme, is lysozyme, this is not produced by bacteria but occurs in human tears and egg white. This enzyme cleaves between N-acetylglucosamine and N-acetylmuramic acid and is used to form protoplasts of sensitive bacterial species.

7 Regulation of metabolism

There is a complicated array of enzyme pathways involved in synthesising the metabolites needed for cell growth; for a cell to grow normally it is essential that the flow of metabolites through the various pathways be under a high degree of control, ensuring that no products are deficient or in excess. This is particularly so for microorganisms since they must respond to much wider fluctuations in environmental conditions than the cells of higher organisms, and reactions essential to cell growth are in a state of dynamic balance with a continuous and tightly controlled flux of metabolites through each pathway.

Microorganisms can use a wide range of organic and inorganic substrates, such as catechols and phenols by *Pseudomonas* species, N_2 by *Azotobacter, Rhizobium* and blue-green algae, long chain alkanes by yeast, sugar alcohols by fungi and starch and other polymers by *Bacillus* species. However, in the presence of more readily assimilated substrates these organisms do not waste energy synthesising enzymes needed to handle these substrates. Cells have the (genetic) potential to carry out a large number of reactions, but control the expression of this potential according to demand.

Apart from the considerable academic interest, there are a number of practical reasons for studying cell regulation. Metabolic pathways can be controlled to suit the demands of research and industry. As an example, the essential amino acid L-lysine is produced quite cheaply by fermentation using overproducing strains which were selected for altered synthesis of methionine and threonine (see p. 104). Some diseases, notably cancer, may be due to changes in the regulation of cell activity. A study of control therefore may indicate possible ways to prevent or treat these diseases. With the present understanding of control in a few bacteriophage and bacterial systems, and current development of techniques for gene transfer between organisms, it is likely that hybrid, microbial systems will be used in commercial production of valuable compounds found only in higher organisms, e.g., hormones, such as insulin.

SITES OF MICROBIAL REGULATION

There are many sites at which the metabolism of an extracellular compound can be controlled; some are indicated in Fig. 7.1.

Entry The cell membrane acts as a barrier to most hydrophilic molecules, but as we have seen in Chapter 3, has systems for transport of specific compounds.

135

prokaryotic

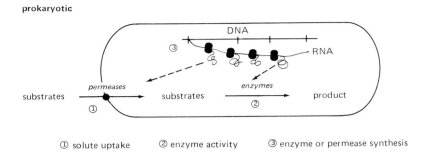

① solute uptake   ② enzyme activity   ③ enzyme or permease synthesis

eukaryotic

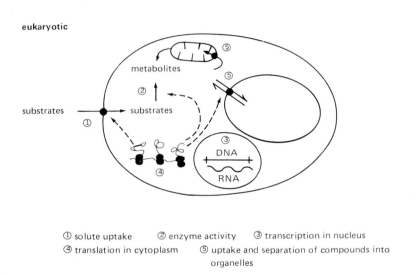

① solute uptake   ② enzyme activity   ③ transcription in nucleus
④ translation in cytoplasm   ⑤ uptake and separation of compounds into
organelles

Figure 7.1   Sites of microbial control

Some of these systems are energy-dependent, and entry can be controlled by the availability of ATP for permease function, or by regulating the synthesis of the permeases themselves.

Flux through the metabolic pathway   In prokaryotes there are two major ways that the flow of metabolites through a sequence of enzyme-catalysed steps is controlled. These are: regulation of the *amount* of enzyme present (by increasing or decreasing the rate of synthesis or degradation of crucial enzymes in the pathway, relative to other cell proteins); and, modifying the *activity* of enzyme molecules already present.

Limiting physical access of enzymes to their substrates   This is more evident in eukaryotes where substrates can exist in separate *pools* by virtue of their location within different membrane-bound organelles. For example, in *Neurospora*, arginine is found in two kinetically distinct pools, one in the cytoplasm, the other in the vacuole, In these two locations the enzymes involved in arginine metabolism occur in very different amounts. This type of control can occur in prokaryotes when the enzymes of a reaction sequence exist as a multienzyme complex, or as membrane-bound entities.

Control as a function of Metabolic Role of a Pathway

The controls which have evolved for metabolic processes depend a great deal on the nature of the substrates and products concerned, and their role in the overall metabolism of the cell. For example, the enzymes of the Embden-Meyerhof pathway, the TCA cycle and anaplerotic pathways supply a flow of compounds, energy and NADPH needed for other biosynthetic processes. Such pathways serve many functions and are usually controlled by modulation of enzyme activity rather than at the level of enzyme synthesis. Not all enzymes in the pathways need be controlled; those occurring just after major branch points in metabolism are most likely to be affected, since they afford the most economical site for control.

For such enzymes, regulation is often complicated, and one enzyme may respond to many metabolites. A good example is citrate synthase which in some species of aerobic Gram-negative bacteria is inhibited by α-oxoglutarate and NADH. NADH inhibition is however reversed by AMP. In other species ATP acts as an inhibitor, while in yet others the substrates, acetyl-CoA and oxaloacetate, stimulate the activity of citrate synthase. These controls reflect the importance to the various organisms of the need to generate ATP and reducing equivalents on the one hand and TCA cycle intermediates for biosynthesis on the other.

For a synthetic pathway with one major end product (for example an amino acid) it is advantageous to a cell if control of a pathway responds to the final concentration of the product in the cell. This is *feedback* control and can be exerted on either enzyme synthesis or enzyme activity, in many cases at both. For a degradative pathway it is more likely that control responds to the presence of the substrate and not its final product. This control is usually found to affect the amount of enzyme synthesised and/or entry of the substrate into the cell.

Control of Enzyme Synthesis

The levels of many enzymes in microorganisms, especially bacteria, are subject to considerable variation depending on the medium in which they are growing.

The most extensively studied control system is that affecting lactose ferment-ation in *E. coli*, and is now fairly well characterised due largely to the work of Jacob and Monod. While the lactose operon is a very useful example for general discussion it should be understood that it is but one of a large number of variations on a theme. There are differences not only from one control system to another, but also between bacterial species in the details of control of a single pathway.

137

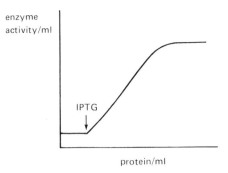

Figure 7.2 Induction of *β-galactosidase* in *E. coli* by the lactose analogue IPTG

LACTOSE OPERON

Induction of enzyme synthesis

Lactose entry into a cell is mediated by a *permease* system. Once inside the cell it is hydrolysed by *β-galactosidase* to glucose and galactose. Neither the permease, nor β-galactosidase is synthesised in large amounts by *E. coli* unless either lactose or certain other β-galactosides are present in the medium. The galactosides *induce* synthesis of the two proteins (and also of a third, *thiogalactoside transacetylase*) from a *basal level* of about 10 molecules per cell to an induced level which may be 1000-fold greater.

Induction occurs quite rapidly after the addition of the inducer and is usually studied using an analogue, *isopropylthiogalactoside* (IPTG) which acts as an inducer, but not as a substrate for β-galactosidase. When studying induction in a growing culture it is important to express the results in a way which shows quite clearly that there has been a change in the *rate* of synthesis of the enzyme relative to the total rate of protein synthesis. One method, used in Fig. 7.2, is to plot enzyme activity per ml of culture against protein per ml of culture. Another method sometimes used is to follow specific activity (enzyme activity per mg protein in culture) as a function of time. On removing inducer, the rate of enzyme synthesis decays rapidly with a half life of about 1 minute. This rapid response in bacteria means that no energy is wasted on unnecessary enzyme synthesis when lactose is exhausted.

The control protein in this system, the *lac* operon, as well as many enzymes in other systems, are not subject to marked changes in the rate of their synthesis by any inducers. The synthesis of these proteins is said to be *constitutive*.

Control System

The operon model for induction of lactose-utilising enzymes is outlined in Fig. 7.3, which illustrates the arrangement of genes determining the various elements of the

138

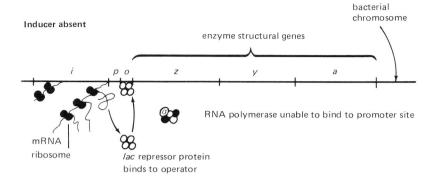

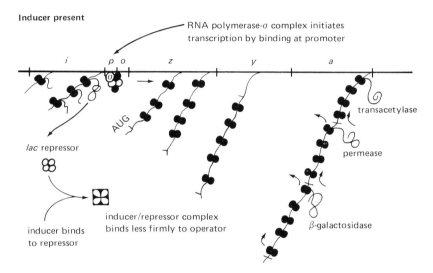

Figure 7.3   The *lac* operon in *E. coli*

lac system and their interaction with the inducer molecule (lactose or β-galactoside analogue) and each other. This model is the culmination of extensive genetic and biochemical study and is based largely on: mapping of mutations affecting lactose fermentation; analysis of control mutants (e.g. constitutive lactose fermenters) by diploid genetics to test the interaction of various elements in the system; and eventual biochemical isolation and purification of all components of the system and their use in *in vitro* systems for synthesising a protein from its DNA template and essential substrates.

The essential features of the model are:

i   The *structural* genes coding for the three proteins, β-galactosidase (*z* gene), permease (*y*) and transacetylase (*a*) are adjacent on the *E. coli* chromosome.

139

ii   The expression of these genes (synthesis of the proteins) is regulated at the level of transcription by control genes which map outside, but close to the structural gene z. One control gene, i, codes for a protein (the repressor protein) which binds specifically and with very high affinity (dissociation constant: 10^{-13} M) to another control site, the operator o, (a short sequence of about 10 nucleotides not coding for a protein). These control genes were initially defined by mutations affecting the inducibility of the system by lactose. In the absence of inducer molecules the i gene product binds to the operator site, o, and *prevents transcription* of the z, y and a genes by RNA polymerase. *In vitro* this complex decays with the incredible half-life of 30 minutes. When inducer is present it binds to the repressor protein as a ligand and the complex so formed has reduced affinity for the operator site. Transcription of z, y and a then occurs from a single binding site for the RNA polymerase—σ factor complex, the promoter region, p. The operator and promoter sites for the *E. coli lac* operon have been sequenced and found to show some degree of overlap. This region involves a total of about 70 nucleotides before the start codon for β-galactosidase synthesis.

iii   The three genes z, y and a are all transcribed as a single mRNA transcript. This polycistronic message contains three initiation sites recognised by *E. coli* ribosomes and acts as template for synthesis of all three proteins. These are therefore produced in a *coordinately* controlled way under normal circumstances, that is, regardless of the level of induction all three proteins are synthesised in the same proportions.

iv   Rapid degradation of mRNA by a specific endonuclease ensures that there is a rapid response in the rate of synthesis of the enzymes on depletion of lactose.

The essential features of an operon are that it consists of a clustering of genes for related function transcribed from a single promoter site under the control of a single operator region.

The *lac* operon is under *negative* control since the repressor protein prevents transcription when bound to the operator. There are other systems, e.g. the arabinose and maltose operons in *E. coli*, part of the galactose fermentation complex in yeast, and certain elements in lytic development of bacteriophage λ (see next chapter) in which the presence of a control protein promotes transcription. This is *positive* control, and evidence available from fungal systems indicates that many control systems in eukaryotic microorganisms show positive control.

Characteristics of Operon Control

A number of important principles can be drawn from studies of operon control in prokaryotes.

i   In addition to structural genes there are control genes. Some of the latter code for proteins binding to DNA with a high specificity to inhibit or promote transcription. Control is therefore largely at the level of transcription and not translation.

ii The DNA-binding activity of control proteins is modified by small molecule ligands (e.g. inducers), in the same way as the activity of allosteric enzymes (see p. 150).

iii Response to a particular molecule in the environment is rapid, due to a rapid alteration in the *rate* of protein synthesis. However, there is no sudden change in the activity of enzyme already present in the culture. This must be diluted out by growth unless it is unstable and degraded by proteases. In bacteria most enzymes do not undergo turnover until the culture is starved.

iv Clustering of related genes can provide a simple way of coordinating the response to a particular environmental change.

OTHER SYSTEMS AFFECTING ENZYME SYNTHESIS IN PROKARYOTES

So far we have seen a brief outline of the *lac* operon, having stressed that it illustrates general principles of prokaryotic control and that other systems differ in detail. These differences often reflect differences in the role of the pathway in question, and the best approach is to compare other bacterial systems. It should however be remembered that most control has been studied in a few enterobacteria which can have the partial diploidy essential to the genetic analysis of control. Other bacteria occupying niches completely different from the human gut may show even more fundamental, as yet undiscovered, differences.

Feedback Control

Lactose utilisation is a catabolic process; many other controlled enzyme pathways are anabolic including those for amino acids, nucleotides, vitamins and cofactors. In these the important variable in overall control is the *end product*, not the substrate of the pathway. Repressor proteins are still formed, but these interact with the end product of the pathway to prevent further enzyme synthesis when end product is in excess. For example, the tryptophan synthesis system in *E. coli* (Fig. 7.4) and the histidine biosynthetic pathway in *Salmonella typhimurium* are negatively controlled operons otherwise analogous to *lac*. An interesting feature of the *his* pathway is that feedback repression appears to respond not only to histidine, but also to various histidyl-tRNAs.

Autogenous Control

The lac repressor appears to be synthesised constitutively. Autogenous systems are those in which the repressor protein is directly involved in the control of its own synthesis. This occurs in the histidine utilisation (*hut*) system of *S. typhimurium* (Fig. 7.5). On theoretical grounds it has been suggested that this type of control can be more stable and responsive to change than that involving constitutive repressor synthesis.

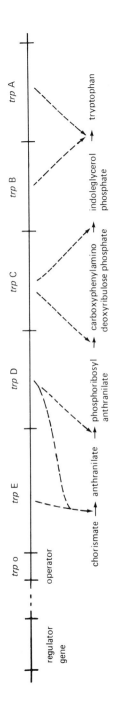

Figure 7.4 The tryptophan synthesis operon in *E. coli*

142

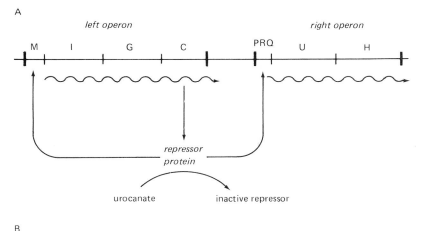

A

left operon   right operon

M   I   G   C   PRQ   U   H

repressor
protein

urocanate   inactive repressor

B

                    H   U   I   G   glutamate
Histidine ⟶ urocanate ⟶ IPA ⟶ FGA ⟵
                                                         formamide

H: histidase   I: 4-imidazolone-5-propionate aminohydrolase   U: urocanase
G: N-formimino-L-glutamate formiminohydrolase   M: promoter on left operon
P,R,Q: promoter/operator complex on right operon   C: repressor protein

Figure 7.5 Control of the histidine utilisation (hut) pathway in *Salmonella typhimurium*; A, Gene arrangement; B, Biochemical pathway

Mixed and Divergent Operons

In some pathways not all enzymes are coded on a single polycistronic message, several operons or an operon and several scattered genes being found. In some of these e.g. the pyrimidine pathway in *E. coli*, regulation is still coordinate (p. 140) for the unclustered genes and the operon; presumably one repressor protein can recognise a number of separate operator sites. This is an important point which may have considerable relevance to both regulation in eukaryotic microorganisms and the phenomenon of catabolite repression which are discussed later.

The arginine pathway in *E. coli* illustrates a case of *divergent* operon control in which the operator region maps *between* two structural genes in a cluster. Transcription of these genes presumably occurs in opposite directions (on different strands of the DNA) but is controlled from the same operator region. A similar situation seems to occur in the eukaryotic system concerned with proline oxidation in *Aspergillus nidulans*.

Sequential Induction and Branched-Chain Pathways

A number of organisms, particularly *Pseudomonas* species, can use a very wide range of organic compounds including phenols and other aromatics as carbon substrate. The pathways for conversion of many of these compounds to succinate

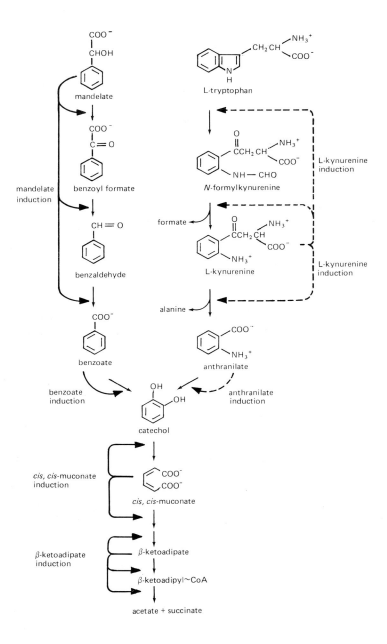

Figure 7.6 Sequential induction of the aromatic degradation pathways in *Pseudomonas*

144

merge (Fig. 7.6), and some substrates are intermediates in the metabolism of others. Such pathways are controlled in sections, rather than as a single operon, so that bacteria can grow on any one of a number of aromatic substrates without inducing unnecessary enzymes. This is accomplished by a process of *sequential induction*, the product of one 'induction block' acting as an inducer for the next group of enzymes in the pathway.

The reverse situation applies when a single intermediate is involved in the synthesis of two or more products by branched pathways. There are several types of control of enzyme synthesis found for branched systems:

i   Feedback repression control on the first enzymes after the last common intermediate (Fig. 7.11).
ii Partial feedback repression by each product on the first enzyme in the overall pathway.
iii Synthesis of different enzymes catalysing the same reaction (*isozymes*) with each isozyme species controlled by one particular end product. This is less commonly found, but a good example is the occurrence of three isozymes for aspartate kinase in *E. coli*: one repressible by lysine, one by methionine, and one jointly by threonine and isoleucine (Fig. 7.11)

Induction, Repression and Operons in Eukaryotes

So far most of the examples discussed have been prokaryotic, and it is interesting to know whether the phenomena of induction and repression, and the organisation of genes of related function into operons, apply to eukaryotes. There are many examples in eukaryotic microorganisms of induction and repression in which the rate of synthesis of an enzyme responds to a particular metabolite. In general, however, the difference between basal and fully induced enzyme levels is usually less for eukaryotes than prokaryotes. For example, *arginase* in *E. coli* can be induced about 100-fold over the basal level, whereas in *Saccharomyces cerevisiae* the maximal change is about 10-fold.

Of more importance is the finding that in those eukaryotes examined in sufficient genetic detail (including *S. cerevisiae, Schizosaccharomyces pombe, Neurospora crassa* and *Aspergillus nidulans*) there are few, if any, examples of significant gene clustering into operons. Usually the structural genes for enzymes concerned with a particular biosynthetic pathway are scattered over many chromosomes and are not linked to each other. For example, enzyme activities for histidine biosynthesis in *Saccharomyces cerevisiae* are coded on six of the seventeen chromosomes. There are a few cases of gene clustering known, the galactose (*gal*) system in *S. cerevisiae* and proline oxidation in *Aspergillus nidulans*, for example. Other clusters occur, although in most cases there has been no distinction between a polycistronic arrangement of different structural genes for different enzymes, and the other possibility of multiple enzyme activities occurring on a single polypeptide coded as a complete cistron.

The *gal* system in *S. cerevisiae* is an operon-like situation with a greater degree of complexity than *lac* in *E. coli*. This can be seen in Fig. 7.7, in which a repressor

145

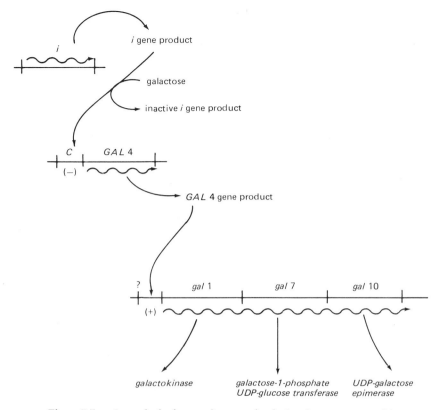

Figure 7.7 Control of galactose fermentation in *Saccharomyces cerevisiae*

protein coded on one chromosome interacts with a control site (*C* gene) on another chromosome in a negative way. This resembles operator/repressor binding. The *C* gene does not however control transcription of the structural genes (*gal* 1, 7 and 10) directly, but regulates synthesis of another control protein (*GAL* 4 gene product) which acts in a positive way to turn on transcription of the three structural genes located on a third chromosome.

So far there is no lack of examples in eukaryotic systems of regulatory genes, in particular *cis*-dominant mutations probably defining operator or promoter-like binding sites. The clustering of related genes into operons rarely applies in eukaryotes but regulator genes can act at separate control sites to coordinate enzyme synthesis.

Histones

Histones are basic proteins found in the nucleus of eukaryotes, but not in bacteria. These proteins bind firmly to DNA and some workdeers feel that they might act as

146

regulators of gene expression in eukaryotes. So far the experimental data are inconclusive in supporting or rejecting this hypothesis. One possibility is that histones control gene expression more permanently during cellular morphogenesis (Chapter 8).

Catabolite Repression

Induction and feedback repression are *specific* responses to a particular metabolite or closely related group of metabolites, e.g. induction of the lactose utilisation system by some β-galactosides. For enzyme systems which metabolise carbon and nitrogen as growth and energy substrates, there are other important controls superimposed on the induction system. These are more general in their action and affect many operons, enabling cells to use a 'preferred' substrate in the presence of a mixture of several others.

The best example of this type of control is the *glucose* effect, illustrated in Fig. 7.8. In the presence of lactose and glucose mixtures, *E. coli* cultures use glucose preferentially, and lactose is not used until the glucose is exhausted. Despite the presence of lactose the *lac* operon is not effectively transcribed until glucose is exhausted. This is one example of several general phenomena, collectively termed *catabolite repression*, in which a number of catabolic reaction pathways are repressed while a preferred substrate (usually, but not always glucose) is available to the cell. Table 7.1 is a summary of some of the enzyme systems subject to catabolite repression and of the organisms in which the phenomenon has been clearly demonstrated. It should be noted that this type of repression is not restricted to the carbon source alone; those enzymes degrading nitrogen-containing

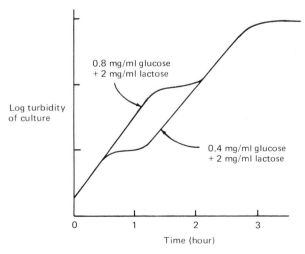

Figure 7.8 Glucose repression of lactose utilisation. The diauxic growth curve

147

Table 7.1 Systems and enzymes subject to catabolite repression

Organism	System or enzyme	Repression exerted by
Escherichia coli	lactose operon 〕 galactose operon 〔 arabinose operon 〔 glycerol kinase 〕	carbon sources, particularly glucose, gluconate and 6-phosphogluconate
	histidine utilisation	carbon and nitrogen sources
	tryptophan utilisation	carbon and nitrogen sources
Bacillus subtilis	sucrase	carbon source
	TCA cycle enzymes	carbon source
	sporulation	carbon and nitrogen sources
Rhizobium species and *Azotobacter*	N_2 fixation	ammonium ions
Klebsiella aerogenes	nitrate reductase	nitrogen source, particularly NH_4^+
Pseudomonas species	glucose oxidation	succinate
Saccharomyces	mitochondrial biogenesis	carbon source
	maltase	carbon source
	arginase	carbon and nitrogen sources
Aspergillus nidulans	proline utilisation	carbon and nitrogen sources
	arginine utilisation	carbon and nitrogen sources
	amidase	nitrogen source

metabolites are also subject to repression by such preferred nitrogen compounds as ammonium ion or glutamine.

The advantages to a microorganism of these repression mechanisms are obvious. If an easily assimilated growth substrate is available the cell does not expend considerable amounts of energy synthesising the enzymes needed for a less efficient pathway, and can use more of its metabolism in producing essential growth components. A good example is *Azotobacter*, in which N_2 fixation is completely repressed in the presence of NH_4^+.

How is catabolite repression mediated at the molecular level? In trying to answer this question it should be stressed that there appear to be a number of different mechanisms operating in those systems which have been carefully studied. So far only one of these, that in *E. coli* involved in glucose repression of enzymes degrading other carbon substrates, is understood in any detail, (Fig. 7.9). This resembles in many ways the control circuit for enzyme induction, with the following salient features:

i   A gene codes for a control protein (CRP, cyclicAMP receptor protein) which binds at specific sites involved in initiation of transcription at *many* operons. Mutants inactive in this gene are unable to induce the synthesis of any catabolite repressible system and can only grow on a rich carbon source such as glucose.

ii   Binding of CRP occurs at or near the promoter site (compare the *lac* repressor binding at the operator). The promoter is also the region specifying correct initiation of RNA polymerase transcription.

148

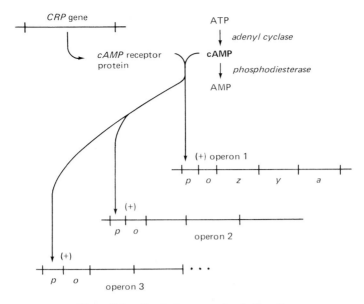

Figure 7.9   Catabolite repression in *E. coli*.

iii   Cyclic $3':5'$ AMP interacts with CRP to promote transcription (note CRP acts positively) and thus relieves catabolite repression. Cyclic AMP and CRP must be added to *in vitro* protein-synthesising systems to obtain maximal activity.

iv   The intracellular level of cAMP in some way reflects the energy status of the cell, probably due to its dependence on the relative activities of the enzymes adenyl cyclase and phosphodiesterase.

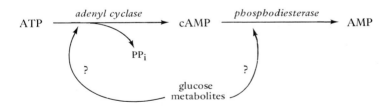

Mutants lacking adenyl cyclase cannot grow on carbon sources subject to repression (e.g. glycerol, lactose, sucrose), but unlike CRP-deficient mutants will respond to cAMP.

This form of catabolite repression is quite general in its effect in *E. coli*, since mutations affecting CRP or adenyl cyclase genes affect most of the glucose repression systems thus far studied. However, cAMP has not yet been detected in some other bacterial species (including *Bacillus*) which are also subject to glucose repression. Some other metabolite may be important in these organisms. Another variation is found in *Pseudomonas* species with metabolism geared to handling a

149

wide range of aromatic compounds producing *succinate*, which represses glucose metabolism in these bacteria.

CONTROL OF ENZYME ACTIVITY

A simple alternative to changing the amount of enzyme present in a cell, is to affect the flux of compounds through a metabolic sequence by altering the activity of one or more *key* enzymes in the pathway. This modulation of activity is usually effected by the reversible binding to an enzyme of a low molecular weight metabolite. The metabolite may be a substrate or product of the sequence, or another compound which reflects the need for that sequence. There are some enzymes whose activities are modified by more than one *effector*, for example phosphofructokinase from *E. coli* has at least three effectors.

In a few cases alteration of activity is brought about by the *covalent modification* of the polypeptide chain(s) of the enzyme. This includes the addition of a low molecular weight group (as in adenylation of *glutamine synthetase* in *E. coli* or the linkage of ADP-ribose to the α-subunit of RNA polymerase during phage T4 infection) or even the specific proteolytic cleavage of a portion of the enzyme molecule (activation of proteases). A third way, very important during morphogenesis, is the complementation of one enzyme with a protein capable of modifying its activity or specificity. There are several examples including the σ factor interaction with RNA polymerase, and in *Saccharomyces cerevisiae* the enzyme ornithine transcarbamylase (catalysing one of the steps in arginine biosynthesis) is inhibited by the arginine-degrading enzyme *arginase.*

Modulation of activity usually provides a fine, continuous control of metabolite flow, and is probably much more responsive to sudden changes or fluctuations in a cell's environment than are induction or repression systems. For the *activity* of an enzyme to alter on sudden increase in repression, pre-existing molecules must be degraded or diluted out. Modulation of enzyme activity is possibly an economical type of control under starvation conditions where *de novo* enzyme synthesis must be accompanied by degradation of other proteins.

Mechanism of Enzyme Modulation

Enzymes under the control of effector molecules have a number of important general properties:

i   They are composed of polypeptide subunits aggregated into multimeric complexes, often tetramers. The subunits can be different, as in the case of the much studied *aspartate transcarbamylase*, or identical. A good example of the latter (and most common) case is the *lac* repressor protein which is not an enzyme, but has many features in common with these enzymes.

ii   The modifiers or effectors are frequently of very different chemical structure to the enzyme substrate, and they bind to the subunit polypeptides at different sites from the catalytic site. In aspartate transcarbamylase, one polypeptide subunit carries the catalytic site, the other, the effector binding site (*allosteric site*).

iii   Studies of these enzymes using a variety of techniques (including NMR, ESR,

150

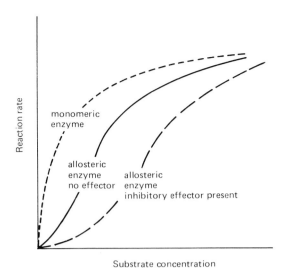

Figure 7.10   Sigmoidal kinetics of allosteric enzymes

circular dichroism and light spectroscopy), as well as chemical modifiers of proteins (such as thiol blocking agents), show that they undergo conformational changes in the presence of the allosteric ligands. This implies that the enzymes can exist in different conformations, hence the name *allosteric* enzymes. The change from one form to another is either brought about by the binding of the ligand inducing the change (induced-fit hypothesis) or by the existence of an equilibrium between the two states with ligand binding leading to the favouring of one of the conformations.
iv   Allosteric enzymes do not show normal Michaelis-Menten kinetics in which the plot of enzyme velocity versus substrate concentation is hyperbolic. Instead, as in Fig. 7.10, the curve is *sigmoidal* and shows a *cooperative* effect at low substrate concentrations; that is, a small increase in substrate concentration leads to a greater than proportional increase in reaction velocity.

It is thus easy to visualise allosteric enzymes as dynamic entities in which the structure of the enzyme is responsive to relatively small changes in the level of both substrate and allosteric activators or inhibitors. These structural changes affect enzyme activity by altering the affinity of the enzyme for its substrate. This differs markedly from the inactivation of enzyme molecules by inhibitors. It alters the reaction rate and enables control to be achieved over a wide range of substrate concentrations.

Patterns of Control

Most reaction pathways are controlled not by just one of the systems we have discussed, but by a combination of many different systems. Rather than discuss this in great detail, it is better to look at one example and remember that there exist

151

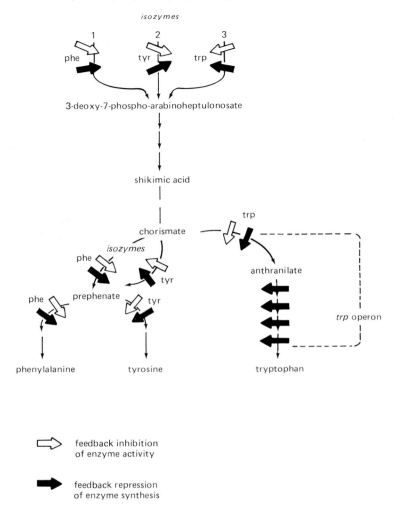

D-erythrose-4-phosphate + phosphoenolpyruvate

isozymes

3-deoxy-7-phospho-arabinoheptulonosate

shikimic acid

chorismate

isozymes

anthranilate

trp operon

prephenate

phenylalanine   tyrosine   tryptophan

⟹ feedback inhibition
of enzyme activity

⟹ feedback repression
of enzyme synthesis

Figure 7.11   Control of aromatic amino acid synthesis in *E. coli*

many variations on the control themes. For different organisms there may be varying demands on a particular pathway, and they will often have different patterns of control. Let us consider the biosynthetic pathways leading to the aromatic amino acids in *E. coli*, outlined in Fig. 7.11. This illustrates:

i   The occurrence of three different isozymes for the first step in the pathway, as a way of regulating the demand from three branches. Each enzyme is controlled by one of the end products either by feedback repression or feedback inhibition, or

152

both. A later step, catalysed by *chorismate mutase* and common to the phenylalanine and tyrosine pathways also has isozymes controlled in the same way.

ii   Feedback repression and inhibition is commonly found just after a branching of the pathways. This enables the cell to avoid an accumulation of intermediates in a pathway when the product of the branch is not required.

iii   The tryptophan branch forms an operon, so that none of the enzymes are synthesised when there is no requirement for their activity. There is however feedback inhibition imposed on this, to modulate the flow of metabolites through this branch when it is functioning.

Finally it should be pointed out that patterns of control can evolve very rapidly, particularly under conditions of high selection. It is not unusual to find that a strain carrying a mutation affecting part of a reaction pathway undergoes mutation at a second, control, site. When branched pathways are involved this can lead to marked changes in cellular metabolism. This has wider implications than its mere nuisance or positive value to the biochemical geneticist; isolates of microorganisms removed from their 'natural' habitat may rapidly undergo heritable changes in their control patterns on subculture under laboratory conditions.

8 Morphogenesis

Morphogenesis is the *development of form.* The term was coined for the intriguing processes whereby a cell or group of cells changes from one shape and function to another. In microbial life cycles there are numerous examples of morphogenetic changes, including at the simplest level the cyclic processes of growth and cell division. Rather than describing many of these life cycles, we will concentrate on the general principles to be learnt from a few systems which have been studied in detail.

GENERAL CHARACTERISTICS OF MORPHOGENETIC SYSTEMS

Events occur as Controlled Sequences

Morphogenesis involves sequences of interrelated biochemical and morphological steps. Of these the morphological changes are most easily studied by electron microscopy; the important biochemical events are often less easy to distinguish. Some, but by no means all of the steps require the expression of genes not used at any other phase of the organism's life cycle, since mutations can be obtained which affect only that one morphogenetic sequence. In microbial development systems there is clearly no change in the *total* genetic information within the developing cell since the final product (for example a spore) often returns to the starting state (the vegatative cell), but there *is* control of the expression this information. In the previous chapter we have discussed the control of metabolic pathways, in this we are primarily concerned with the question of the control of the *timing* of sequential events.

Assembly

During cellular development new structures are formed on old foundations; this requires the assembly of subunit molecules into the correct spatial relationships relative to each other, and raises many interesting questions. How do molecules aggregate correctly to form larger structures such as cell walls and flagella? Is this process of assembly spontaneous or controlled in some way? How are sequences introduced into assembly involving many components? What determines shape?

154

Some microbial differentiation systems involve only changes within a single cell (e.g. bacterial and yeast sporulation, spore germination), but others, particularly those concerned with sexual fusion and/or fruiting body formation depend on extensive cooperation between cells. These often involve initial *communication* and *recognition* stages, such as must occur in zygospore formation in *Mucor* where aerial hyphae of different mating type grow toward each other along the most direct path. In some organisms, once aggregation has occurred, the individual cells do not fuse but retain some integrity (as in cellular slime moulds), while in others surface adsorption is followed by fusion of the cells. In eukaryotes this fusion leads initially to heterokaryons (cells containing nuclei from different parents) and in some cases nuclear fusion also occurs to form heterozygotes. Here one can ask: How do cells communicate at a distance? What confers specificity on cellular recognition? How do fusing cells achieve synchrony with each other?

Clearly, the study of morphogenesis needs an analysis of biochemical, cytological and genetic control processes, together with an understanding of how these interact with each other in function, time and space. When choosing an experimental system there are several desirable features. Biochemical studies on microorganisms require large numbers of cells, hence synchrony of the morphogenetic process is essential. In order to study interactions in biology it is common to interfere with the system in some way. This can be done using inhibitors, or more specifically by *mutation*. The value of genetics for study of control is obvious if one considers the progress made in understanding operon control. This is particularly true if both diploid and haploid states are available; haploids can be used to isolate recessive mutations and diploids can be used to test dominance relationships from which sensible statements about control can be made.

SEQUENTIAL EVENTS DURING MORPHOGENESIS

Morphological Sequences

Bacterial sporulation is the most extensively studied example of cellular development in a microbial system, occurring in the genera *Bacillus, Clostridium* and *Sporosarcina* in vegetative cells facing starvation for either carbon or nitrogen substrates. The spore differs markedly from vegetative cells in shape and structure, as well as in being extremely resistant to heat, radiation and most chemical disinfectants. The changes seen by electron microscopy during sporulation in *Bacillus* are outlined in Fig. 8.1. The process begins with the formation of an axial chromatin filament (*Stage I*), followed by an acentric membrane invagination (*Stage II*). Note that cell wall material is not formed between the now separate prespore and the mother cell. *Stage III* is characterised by engulfment of the prespore by the mother cell membrane leading to inclusion of the prespore in the mother cell cytoplasm. This prespore is now surrounded by two membranes which

L

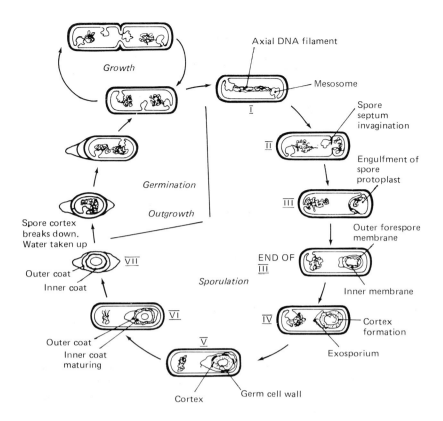

Axial DNA filament

Growth

Mesosome

I

Spore septum invagination

II

Engulfment of spore protoplast

III

Germination

Outgrowth

Spore cortex breaks down. Water taken up

Outer forespore membrane

END OF III

Inner membrane

Outer coat

VII

Sporulation

Inner coat

VI

IV

Cortex formation

V

Exosporium

Outer coat

Inner coat maturing

Cortex

Germ cell wall

Figure 8.1 Morphological stages during sporulation and germination in *Bacillus* species (From *Critical Reviews in Microbiology*, **1**, 479, 1972; by permission of the Chemical Rubber Co.)

have *opposite polarity*. An electron-transparent cortical material is next synthesised between these membranes (*Stage IV*), and if one considers the polarity of the membranes it is apparent that the cortical material is laid down on the 'out' side of these (i.e. that facing into the medium before invagination took place). Not surprisingly the cortex is composed of a modified peptidoglycan. Once the cortex is almost complete a lysine- and cystine-rich protein condenses in lamellae around the outside of the spore (*Stage V*) to form the coat. The spore subsequently matures to a very refractile, sometimes wrinkled structure (*Stage VI*; Fig. 8.2) and is released by lysis of the mother cell (*Stage VII*).

Biochemical Events Also Follow Ordered Sequences

Many biochemical changes have been detected during bacterial sporulation; some of these are summarised in Table 8.1. In many cases however, the relevance of these

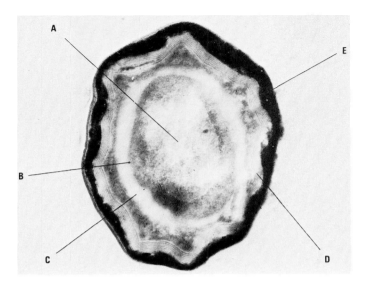

Figure 8.2   Thin section of a mature *Bacillus subtilis* spore; × 3000. A: DNA region; B: Protoplast membrane; C: Cortex; D: Inner spore coat; E: Outer spore coat

events to spore formation is not known. Any bacterial culture which is allowed to approach starvation will undergo numerous changes brought about by the relief of catabolite repression. These need not be necessary to sporulation. There are however a few biochemical changes *specific* to spore formation, and these are useful 'marker' events in studying the process. Two compounds, *dipicolinic acid* (DPA) and the *lactam of muramic acid* have so far been found only in bacterial spores.

dipicolinic acid   muramyl lactam

DPA chelates with calcium ions and forms about 10% of the spore dry weight; it may be involved in heat resistance or maintaining dormancy. The muramyl lactam is a component of the cortical peptidoglycan, which also differs from the cell wall form in being less crosslinked. Another component of the spore, the *coat protein*, is antigenically distinct from proteins found in vegetative cells.

157

Table 8.1 Some biochemical events occurring during sporulation
in *Bacillus* species

Synthesis of:	Morphological stage for onset
proteases	0
antibiotic(s)	0
esterase	0
TCA cycle enzymes	0–I
arginase	0–I
N-succinylglutamate	I
alanine dehydrogenase	II
alkaline phosphatase	III
spore coat protein	III or IV
glucose dehydrogenase	IV
sulpholactic acid	IV
ornithine transcarbamylase	IV
cortex synthesising enzymes	IV
Ca^{++} incorporation into spore	IV
cysteine incorporation (coats)	V
lytic enzyme(s)	VI

The biochemical events always associated with sporulation occur in an ordered
sequence which correlates with that for morphological changes, as indicated in
Table 8.1.

Later Events are Dependent on Successful Completion of Earlier Ones

Studies with asporogenous mutants (those unable to form heat resistant spores)
have shown that the morphological and the biochemical sequences are tightly
coupled to each other, since a mutation blocking sporulation at a particular
morphological stage also prevents all biochemical changes normally occurring after
that stage. For example, *alkaline phosphatase* in *Bacillus subtilis* begins to appear
towards the end of stage II. No mutants blocked prior to stage II produce the
enzyme, but any which begin forespore invagination do.

These results also indicate that many events are *dependent* on the successful
completion of all previous ones. There are some exceptions to this, since spore coat
protein (which assembles on the developing spore at stage V) can accumulate in
cells blocked at stage II. Moreover, once septation occurs there are two chromosomes in the different protoplasts within the cell; both are active in protein
synthesis and appear to be involved in separate sequences of events. These findings
in bacterial sporulation have been confirmed in many other morphogenetic systems,
including bacterial and yeast cell division cycles (p. 39) in which we have already
seen that events are controlled as a pattern of branched, interrelated sequences.

The cellular slime mould *Dictyostelium discoideum* provides an interesting
insight into the strict correlation between morphology and biochemistry. The life
cycle of this organism is shown in Fig. 8.3; amoeboid vegetative cells undergo

158

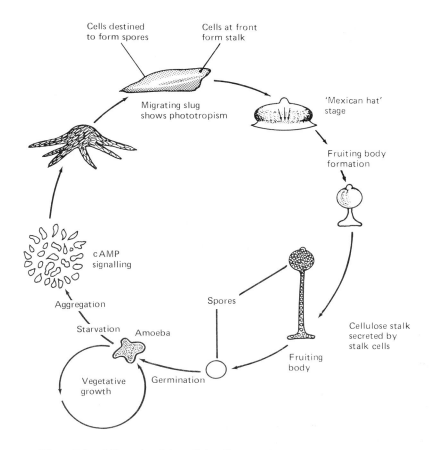

Labels within figure:

Cells destined to form spores

Cells at front form stalk

Migrating slug shows phototropism

'Mexican hat' stage

Fruiting body formation

cAMP signalling

Aggregation

Starvation

Amoeba

Spores

Cellulose stalk secreted by stalk cells

Vegetative growth

Germination

Fruiting body

Figure 8.3   Life cycle of the cellular slime mould *Dictyostelium discoideum*

aggregation and subsequent multicellular stages following a period of starvation. During the programme leading to formation of the fruiting body with differentiated stalk cells and spore cells, a number of enzymes are synthesised following a given sequence. If at the Mexican hat stage the multicellular mass is disaggregated, the amoebae reaggregate, and go through the usual sequence to produce *twice* the normal levels of these enzymes.

Control of Sequences

Initiation

Sporulation in bacteria, yeast, and other fungal systems is usually induced by starvation for either or both C and N sources. It is not a general response to the cessation of growth, only to conditions leading to relief of catabolite repression.

159

There are other examples in fungi in which sporulation is a specific response to CO_2 or light.

Spore germination on the other hand is usually triggered by the presence of one or more particular compounds; for example some *Bacillus* spores are germinated in the presence of L-alanine, others by glucose, nitrate mixtures and still others by K^+, fructose and glucose mixtures, or even Ca^{2+} dipicolinate.

Timing of Sequences

Hypotheses to explain the timing of enzyme synthesis during development in *cellular* systems are quite speculative. In contrast there are several examples of bacteriophage development which have been characterised extensively.

Single Stranded RNA Phages (R17, MS2, f2)   These are very simple indeed, their RNA coding for only three proteins: a *replicase enzyme, a coat* protein and a *maturation* protein. These phage infect *E. coli* by attaching to sex pili, releasing their RNA through the pilus to the cytoplasm. Following infection the three proteins do not appear in the same proportion, nor are they synthesised simultaneously. Moreover, during the late phases of infection there is a marked reduction in the rate of replicase gene translation.

By *in vitro* protein synthesis studies, from the use of mutants affecting each phage gene, and by studying the interaction of the purified gene products with phage RNA it has been found that:

i   The coat protein acts as a translational repressor of replicase synthesis, by binding to the start region of the replicase gene.
ii   The replicase gene cannot be translated until part of the coat gene is translated. This may be due to the secondary structure of the phage RNA, which shows extensive intramolecular base pairing, hiding the initiation site for the replicase in the coat gene.

The important point here is that gene expression is controlled at the level of *translation*, regulating the amount of gene products synthesised (for R17 the ratio of coat protein: replicase: maturation protein is 20:5:1) rather than the timing of their synthesis. Translational repression can however account for some change in the rate of synthesis of a protein at a given time during development.

Bacteriophage λ: A much deeper insight into genetic regulation of phage development comes from studies of the double-stranded temperate DNA phage λ. Its DNA usually integrates into the chromosome of its host, *E. coli*, at a specific attachment site. The λ lysogens so formed are immune to further infection by other phages including some unrelated to λ; they grow normally and the phage genome replicates along with the host genome. Control in this system is concerned with the maintenance of lysogeny and timing of the less frequent lytic development.

During *lytic* development in λ, the λ DNA is excised from the host chromosome and the previously unread genes coding for phage proteins are expressed. These phage-coded proteins are not all synthesised simultaneously. Some appear early

160

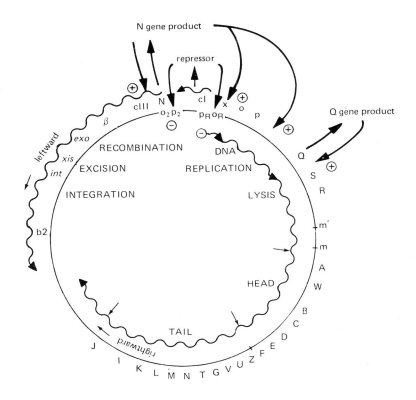

Figure 8.4   Genetic and control map of bacteriophage λ in circular form. RNA transcription indicated by wavy lines; control circuits by solid lines

after the induction of lytic development, while others are produced in significant amounts only at later stages.

What control systems coordinate protein synthesis in this way? In trying to answer this question we should begin by considering the arrangement of genes on the λ genome, shown in Fig. 8.4. For our purposes it is not necessary to know the function of every gene; what should be noticed is that genes coding for related functions (such as head synthesis, tail synthesis, recombination, and DNA synthesis) are clustered into groups. Gene clustering, as we have already seen (p. 140) is one of the main characteristics of operon control, and in the following genetic 'control circuit' it should be possible to distinguish four interrelated operon-like systems. The following control activities have been identified:

cI gene: The product of this gene is the *repressor protein* which acts negatively at two operator sites (o_L and o_R) located on opposite strands of the phage genome, preventing reading of the remaining phage genes and thereby stopping lytic development.

161

o_L, o_R, p_L, p_R When the activity of the repressor is decreased in some way (e.g. by UV treatment) transcription starts at two promoters. From the leftward promoter, o_L transcription proceeds leftward on the 'left' DNA strand, and in the early stages stops just after gene N. This termination appears to be brought about by the *E. coli* ρ factor (see Chapter 6). Similarly, rightward transcription from the o_R, p_R region occurs on the 'right' strand, and terminates between the x and y regions after reading about 1% of the genome.

N gene product   This most important control element acts at a number of sites to control the *timing* of the whole development. It is a positive control protein with 'antitermination' activity, permitting leftward transcription beyond N to continue indefinitely. This produces the proteins concerned with prophage excision and recombination. Similarly, it acts on rightward transcription allowing unhindered reading of genes O and P, and acts again later, beyond P at a termination site just before gene Q.

O, P gene products   These proteins modify the host DNA polymerase so that λ DNA can be replicated, increasing the number of gene copies available for transcription, and amplifying the number of head and tail proteins made.

Q gene product   This protein in some way increases the transcription of late genes 'downstream' on the right strand by a factor of about ten. Head and tail proteins are thus produced in large amounts as late events in lytic development. The exact nature of this protein is not known but it may act as a phage-coded σ factor (see p. 124) which complements the *E. coli* RNA polymerase enabling it to recognise a promoter after Q.

We can now summarise the important concepts which arise from studies with λ:

i   Some proteins produced during phage development act as regulators controlling the timing of gene expression (e.g. *cI*, N and Q). These controls operate mainly at the level of transcription.
ii   The arrangement of genes into sequences of operons is one way of coordinating the synthesis of proteins involved in a particular step of development.
iii   Gene amplification can increase the production of proteins to levels needed for successful assembly.
iv   Although control of the timing of protein synthesis is largely at transcription, there is probably some translational control similar to that found in R17. Thus the λ head and tail proteins, translated from the same message, are not produced in equimolecular proportions.

Biochemical Aspects of Sequential Control

So far we have discussed genetic control circuits. How, in chemical terms does the cell switch from using one set of information to another? One answer to this question was found not in λ, but in the lytic *E. coli* phage T4. The T-even phages

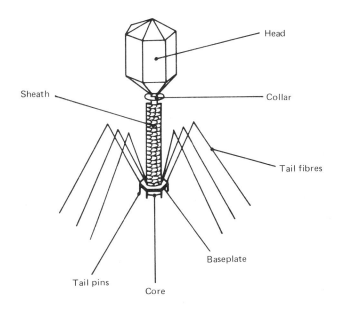

Figure 8.5   Structure of bacteriophage T4

have a complicated structure shown in Fig. 8.5; many genes are involved in the assembly and phage-coded proteins appear sequentially during the lytic development. Very soon after infection the synthesis of bacterial proteins is shut off, and only T4 genes are read. This switching is achieved quite simply: after infection only a few phage genes (with bacterial-like promoters) are open to transcription by the *E. coli* RNA polymerase, most are not expressed. One of these 'early' T4 genes is a new σ factor which displaces the host cell σ factor from the RNA polymerase complex, see Fig. 8.6. In addition a second early phage gene codes for an enzyme which modifies the α subunit of the host RNA polymerase core enzyme by adding a diphosphoadenosyl group. These two changes prevent the RNA polymerase complex from initiating transcription at bacterial promoters, and only late phage genes can then be read.

Timing Control in Cellular Development

Modification of RNA polymerase is also important during bacterial sporulation. This was first indicated by the inability of lytic phage φe to develop in *sporulating* cells of *Bacillus subtilis*, due to an alteration in the β subunit of RNA polymerase such that the enzyme could no longer be complemented by the vegetative σ factor. The β subunit (mol. wt. 155 000) is probably specifically cleaved by an *intracellular* protease to a lower molecular weight form (110 000). It is not yet clear when this change occurs, nor whether any new σ factors are formed during sporulation, but RNA polymerase is somehow involved in control since some mutations

163

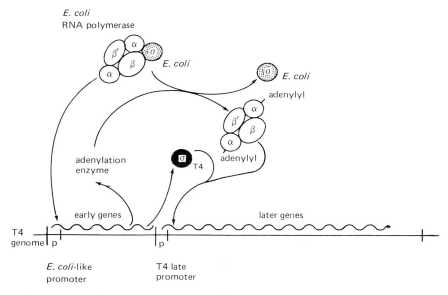

Figure 8.6 Switching of gene transcription by phage-induced modification of *E. coli* RNA polymerase after T4 infection

specifically affecting its β subunit (selected for resistance to the antibiotic rifamycin) also prevent sporulation in *B. subtilis* without affecting vegetative growth. It seems unlikely that sufficient modifications could occur to the RNA polymerase core to account for the timing during sporulation, and the above may be just one major switching element preventing transcription of unwanted vegetative genes and opening up sporulation genes.

Unlike the situation in λ, sporulation genes are scattered in at least thirty groups widely separated on the chromosome in random order in terms of the stage at which they act. Timing cannot therefore be explained in terms of *sequential transcription*. So far there are few clues as to how timing is introduced. One hypothesis suggests a process of *sequential induction* in which one product of a group of genes switched on as an operon or set of operons, acts as an inducer for the next block of genes, as indicated in Fig. 8.7. Notice that this differs little from the hypothesis that new σ factors are formed for each block of genes since both inducers (or repressors) and σ factors are proteins acting at the operator-promoter region to regulate transcription. There are other possibilities, and some switching may occur at the level of translation, controlled by the relative affinities of mRNA binding sites for ribosome subunits.

ASSEMBLY

Growth of cell surfaces involves the site-specific deposition of new wall material (Chapter 2), and the control of assembly at this level is probably

164

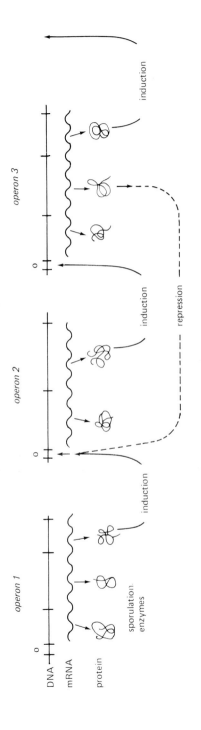

Figure 8.7   Sequential induction model proposed to account for timing of gene reading during bacterial sporulation

165

quite complicated. More is known, however, about the genesis of smaller and more regular structures such as pili, fimbriae and flagella. However, once again viruses have been most extensively studied and can be used to illustrate fundamental concepts of assembly due to their simplicity of composition and regularity of shape.

Viral Assembly

Viruses can be grouped in terms of their shape. Thus animal and plant viruses can appear roughly spherical or filamentous in the electron microscope, while bacteriophages can be either spherical or have the more complicated head and tail arrangement exemplified by the T-even phages (Fig. 8.5).

Electron microscopic examination of simpler spherical viruses has shown that their coats, which surround and protect the nucleic acid, are composed of many apparently identical subunits. This illustrates an important principle of *morphopoiesis* (a term used for the process of assembly): *large structures are constructed most efficiently by repetitive self-assembly of identical subunits.* This efficiency is seen as a reduced need for genes to specify a large structure, and is clearly essential in the case of viruses with a very limited coding potential. The subunits, or *capsomeres*, are arranged in a particular type of cubic symmetry, icosahedral symmetry (an icosahedron has twenty faces, twelve vertices and thirty sides) which is the minimal energy arrangement of a closed shell built of regularly-bonded and identical subunits. That is, it approximates a sphere and produces a very strong structure. The subunit clusters in icosahedral viruses are arranged either as hexamers or pentamers, pentamers occurring only at vertices (there are only twelve in each spherical particle). If subunits can bond to each other by at least two different bonds they can associate into hexagonal array. Slight distortion of this bonding produces pentamers and forms a vertex rather than a flat sheet structure, as illustrated in Fig. 8.8. This introduces the principle of *quasiequivalence* which implies that binding between subunits can be distorted non-randomly to give a more stable structure. When the phage coat (or *capsid*) of a spherical virus is composed of more than sixty subunits then quasiequivalent bonding is necessary to provide both pentamers for vertices and hexamers between the vertices.

Filamentous viruses are formed by helical arrangement of identical subunits; helices are the natural consequence of joining subunits by a single bond arrangement. This arrangement is found in viruses such as tobacco mosaic virus, in the tails of more complicated bacteriophages; and in bacterial flagella and pili.

Control of Assembly

Is subunit assembly controlled? The simplest possibility is that once formed, subunits undergo automatic self-assembly into the final structure. This is probably true for simpler viruses, and *in vitro* assembly of capsid structures has been achieved using dissociated capsomeres. For more complicated phage built up from several

166

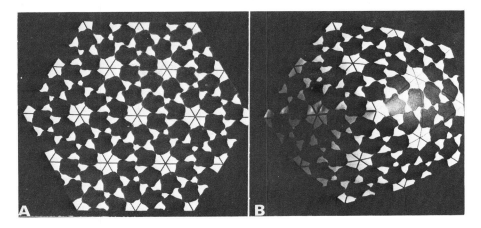

Figure 8.8 Quasiequivalence and vertex formation in icosahedral assembly. A and B show vertex formation by introducing slight distortion in a flat hexagonal array of subunits leading to a pentamer

different structures the picture is more complicated since sequences are introduced into the assembly line.

Morphopoiesis of T4 has been the most extensively studied. A large number of phage genes are involved in the assembly, but by no means all of them code for polypeptides appearing in the completed particle. Some presumably exert control over and above that of direct self-assembly. For example, eight genes are known to be involved in the assembly of the T4 head, although the final head is composed largely of a single protein species arranged into an elongated version of an icosahedron. Other genes ensure correct assembly and mutations in these lead to aberrant development, including the production of elongated heads, short heads or heads unable to package DNA. Initially a procapsid is formed by assembly of gene 23 product on a core protein, involving the participation of shape-specifying proteins. Subsequently a phage-coded protease specifically cleaves the gene 23 polypeptide inducing a conformation change in the major subunit. At the same time other proteins complex with the procapsid and phage DNA is taken up. Here again we see how the modification of proteins by specific cleavage can be important in controlling morphogenesis.

INTERCELLULAR INTERACTIONS

Cooperation between cells during morphogenesis is seen most clearly in eukaryotic microorganisms; there are fewer examples in bacteria and blue-green algae. However in one group of bacteria, the myxobacteria, there is extensive interaction between cells during the development of multicellular fruiting bodies. The process in *Myxococcus* is typical, and is shown in Fig. 8.9. In the presence of an adequate

167

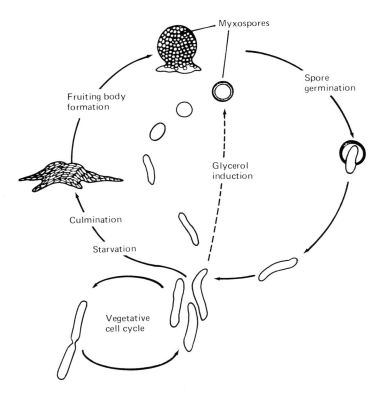

Figure 8.9   Life cycle of *Myxococcus xanthus*

supply of nutrients these bacteria multiply, and glide over the surface of the substrate. Even at this vegetative stage individual cells are attracted to the colony from which they are derived. Cells at the edge of the colony can move away, but return eventually or stop until the colony catches up, so that the whole colony moves as a loose, dynamic conglomerate. When nutrients are exhausted the organisms aggregate into a mound, and eventually form up into a more or less complex fruiting structure characteristic of each species. In *Myxococcus xanthus* this structure is relatively less differentiated, forming a single spherical structure, while in *Stigmatella* the fruiting body comprises a stalk from the top of which radiate macrocysts; within these the vegetative cells develop into cysts (or myxospores).

This illustrates some of the important elements of intercellular interaction: *communication* between cells to maintain contact; *recognition* associated with this communication and the later stages of aggregation into the fruiting body. In the following discussion, in moving from description to consider the physiology of interactions between cells we will concentrate on those systems most extensively studied. These are mainly eukaryotic but again we find bacteriophages provide useful models.

168

Communication and Recognition

Recognition at Cell Surfaces

In liquid suspension interacting cells can make contact, even if non-motile, by a process of random diffusion and collision. This occurs in algal and yeast mating (in which two cells of opposite mating-type recognise each other and fuse), mating during bacterial conjugation (recognition between sex pilus and recipient cell surface) and during bacteriophage adsorption to its host. In each system recognition occurs at the cell surface.

Bacteriophage adsorption to the host illustrates clearly how this recognition occurs. Most phages show a very restricted *host range*, often infecting only a few strains of one species. Moreover, different phages adsorb to different surface structures of the host, sex pili, flagella, or bacterial surface polymers. In the case of phages attaching to the host surface lipopolysaccharide (including T4 and *Salmonella* phage P22) recognition involves a protein of the phage tail-fibre binding to a *specific* sequence of sugar residues in the lipopolysaccharide. This is probably very similar to enzyme-substrate interactions — such as that between lysozyme and its peptidoglycan substrate in which six sugar residues lie within a long groove

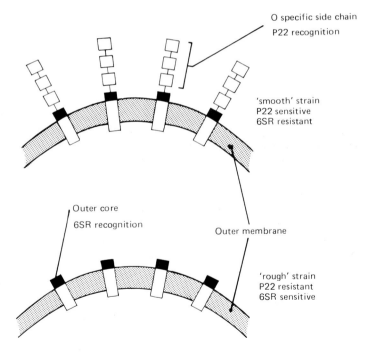

Figure 8.10 Host specificity of *Salmonella* phage P22 and 6SR and the nature of the lipopolysaccharide in the host outer membrane

of the lysozyme molecule, or antigen-antibody recognition. The specificity is illustrated by the host ranges of *Salmonella typhimurium* phages against various bacterial mutants blocked at different stages of completion of lipopolysaccharide. For example P22 recognises the O-specific side chains only while 6SR adsorbs to the outer core region only when it is exposed in the absence of the O-antigen (Fig. 8.10).

This principle of surface interaction between complementary molecules extends to other, higher systems. In mating in yeast, recognition between opposite mating types involves agglutinins which are *mannan-protein* complexes in *Hansenula wingei*, and in the alga *Chlamydomonas* recognition between (+) and (−) gametes involves a sugar sequence of a glycoprotein present on the tip of flagella of (+) strains and probably a protein conformation on the (−) gamete flagella.

Communication at a Distance

When cells interact at an interface, or in the atmosphere above a substrate, there is a need for communication and/or recognition over a distance. This is seen clearly in two systems: in *Dictyostelium discoideum* during aggregation of amoebae prior to the multicellular phases of fruiting body formation, and in the formation of *zygospores* in *Mucor* and related genera.

In the cellular slime moulds, amoebae multiply singly, but begin to aggregate when they reach starvation. This occurs at surfaces, and is a chemotactic response to a low molecular weight compound (*acrasin*) identified as cyclic AMP which acts as a messenger molecule across the whole spectrum of biological complexity (e.g. see discussion on catabolite repression, p. 147). Slime mould aggregation has been extensively studied as a model for intercellular communication, and so far the findings can be summarised:

i   Soon after the onset of starvation 'pacemaker' cells secrete cAMP in pulses.
ii   All amoebae have a receptor protein for cAMP located in the surface membrane, and in addition a potent phosphodiesterase that hydrolyses cAMP.
iii   Cells stimulated by a pulse of cAMP move briefly toward the source of the pulse and then secrete a pulse of cAMP themselves. This leads to waves of inward movement (Fig. 8.11) which are gradually changed into radial 'streams' of cells moving towards aggregation centres.
iv   Each pulse of cAMP elicits profound changes in the shape and cohesive properties of the amoebae, leading to strong end-to-end contact.

So far it is not clear how the reception of a cAMP pulse brings about shape changes or even the pulse relay system. What is clear however is that this communication system (and probably that in the myxobacteria which also display wave-like propagation of signals) depends on the synthesis by cells of a diffusible molecule which can be detected, relayed *and broken down* by the receiving cells.

Zygospore formation in *Mucor* occurs above the surface of the growth medium; (+) and (−) hyphae in proximity are induced to produce specialised hyphae, *zygophores,* which grow towards each other with a high degree of recognition of

170

both direction and mating type, illustrated in Fig. 8.12. This zygotropism is presumed to be mediated by the synthesis of distinctive molecules by each type of zygophore; these must diffuse in the vapour phase, and be detected by the zygophore of opposite mating type to elicit the correct growth response.

The presence of a single chemical, *trisporic acid,* controls the induction of zygophores in unmated cultures of either (+) or (−) strains. This 'hormone' acts at very low concentration ($<10^{-8}$ M). How can the production of a single compound elicit the specific response of both (+) and (−) strains when neither alone can synthesise it, but both together can? One suggestion is that each mating type produces intermediates in trisporic acid synthesis up to a particular step. To complete formation of the hormone, two reactions are needed, and *these can occur in any order.* When close to each other, both mating types can complement each other in the complete synthesis of trisporic acid:

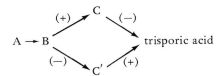

C is produced in (+) strain alone, C′ in (−) strain alone.

It may be that trisporic acid is involved in zygotropism as well as zygophore induction. If not, volatile hormones which are complementary and specific to both transmitters and receivers must be involved together with a mechanism for chemoreception and for transducing the reception to a directed growth response. Many intriguing problems remain to be solved.

Phasing of Cells

Another form of communication found during mating is exemplified by the mating hormones in *Saccharomyces cerevisiae.* Haploid cells of opposite mating type (a and α) fuse to form heterokaryons briefly, followed by heterozygosis. If two cells were not at the same stage of the cell cycle, then fusion would present problems. However α cells excrete a diffusible protein factor into the medium which arrests *a* cells at the beginning of the cell cycle, and similarly *a* cells produce a factor which has similar effects on α cells. The α factor also induces temporary shape changes in *a* cells, causing an elongation of one 'end' of the ellipsoidal cells, which may be essential for the mating process.

Fusion of Cells

Cell fusion occurs during mating in eukaryotic microorganisms following the recognition and surface adsorption processes. In fungi and algae with rigid cell walls there is a need for specific breakdown and in some cases resynthesis at the site of contact. Once wall breakdown has been accomplished membrane fusion appears to

Figure 8.11   Outward wave propagation from pacemaker cells in aggregating *Dictyostelium* amoebae. (Courtesy of Dr. Marilyn Monk)

occur spontaneously, since it can be induced in some organisms by mixing of protoplasts.

There are many aspects of morphogenesis we have not touched upon — for example the cooperation of cells in fruiting structures so that some form spores, others form stalks (as in *Dictyostelium*) — in which cell position and even cell 'history' may play important roles. In general less is know about these phenomena and space does not permit extensive discussion; there are however a number of very interesting reviews listed below.

The points we have discussed are general to developmental biology and are not confined to microbiology. Ultimately one wants to be able to find answers to such questions as: 'how does a fertilised egg give rise to all the various cell types found in the mammalian body?' This is obviously extremely complicated and micro-organisms provide a way of analysing development in simpler systems to formulate principles testable in higher organisms.

172

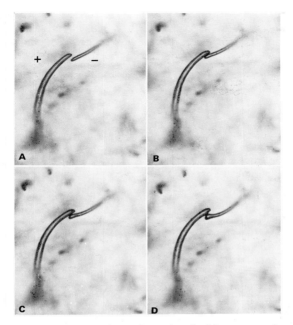

Figure 8.12 Zygophore formation in *Mucor mucedo*
A: The two zygophores attracted by zygotropism; B: 4
minutes later; C: Contact, 1 minute later; D: Swelling at
point of contact, 10 minutes later (Courtesy Dr. G. W.
Gooday)

   This is however not the sole justification for this type of research. Microbial
development is intrinsically a fascinating subject and worth study for its own sake.
For those seeking economic reasons, bacterial spore formation and germination are
of considerable importance in food processing and other sterilisation industries due
to the extreme resistance of spores to desiccation, heat, radiation and disinfectants.
Moreover for many fungal pathogens of both plants and animals morphogenetic
processes seem to play a role in pathogenicity, exemplified by the yeast to
pseudomycelial transition in *Candida* species.

Further reading

General sources of review articles covering most aspects of microbial physiology are given in the introduction.

Chapter 1

MANDELSTAM, J. & McQUILLEN, K. (1973) *Biochemistry of Bacterial Growth*. Blackwell Scientific Publications, Oxford. The major bacterial cell structures are covered in some detail in Chapter 1.

BERG, H. C. (1975) Bacterial behaviour, *Nature,* **254**, 389–392. A very interesting discussion of how bacteria move and respond to chemical stimuli.

BROCK, T. D. (1970) *Biology of Microorganisms*. Prentice-Hall, New Jersey. Gives a rather more extensive account of structures than there is space for here.

ROGERS, H. J. (1970) Bacterial growth and the cell envelope. *Bacteriological Reviews,* **34**, 194–214. Good account of cell wall structures.

Chapter 2

DONACHIE, W. D., JONES, N. C. & TEATHER, R. (1973) The bacterial cell cycle. *Symposium of the Society for General Microbiology,* **23**, 9–44. A clear outline of what is known, and various hypotheses, for the control of the cell cycle in bacteria.

HARTWELL, L. L. (1974) *Saccharomyces cerevisiae* cell cycle. *Bacteriological Reviews,* **38**, 164–198. An extensive treatment of the cell cycle in this eukaryote. The article also provides much useful information on the molecular biology of microbial eukaryotes.

MITCHISON, J. M. (1970) *The Cell Cycle*. A very readable synthesis of the literature on the cell cycle.

HUGO, W. B. (ed.) (1971) *Inhibition and Destruction of the Microbial Cell*. Academic Press, London and New York.

Chapter 3

HAROLD, F. C. (1972) Conservation and transformation of energy by bacterial membranes. *Bacteriological Reviews* **36**, 172–230. A critical and interesting review of facts and hypotheses concerning the energetics of transport across mitochondrial and bacterial membranes.

KENNEDY, E. P. (1970) The lactose permease system of *Escherichia coli*. In *The Lactose Operon* (eds J. R. Beckwith & D. Zipser) 49–92. Cold Spring Harbor Laboratory. The transport system studied in considerable detail chemically and genetically.

Chapter 4

LEHNINGER, A. L. (1971) *Bioenergetics*. Benjamin, Menlo Park, Useful source for biochemical background.

PECK, H. D. jun. (1968) Energy-coupling mechanisms in chemolithotrophic bacteria. *Annual Reviews of Microbiology*, **22**, 489–518. Covers the chemical transformations producing energy in this group of autotrophs.

PFENNIG, N. (1967) Photosynthetic bacteria. *Annual Reviews of Microbiology*, **21**, 285–8. An interesting review of the physiology of this group.

ROSE, A. H. (1968) *Chemical Microbiology*. 2nd ed. Butterworth, London. Elaborates in some detail many different pathways for energy production found in a wide range of microorganisms.

Chapter 5

COHEN, G. N. (1967) *Biosynthesis of Small Molecules*. Harper & Row, New York, Evanston and London. Brief and clear outline of the major biosynthetic pathways.

MANDELSTAM, J. & McQUILLEN, K. (1973) *Biochemistry of Bacterial Growth*. Blackwell Scientific Publications, Oxford. Chapters 3 and 4 give a detailed picture not only of the biosynthetic pathways in bacteria, but also of the way they are regulated. The best single reference source.

Chapter 6

British Medical Bulletin, **29**, 3, September 1973. Advances in molecular genetics. A collection of articles covering replication, transcription and translation in prokaryotes and eukaryotes.

CLARK, B. F. C. (1974) Protein Biosynthesis. In *Companion to Biochemistry*, eds A. T. Bull, J. R. Lagnado, J. O. Thomas & K. F Tipton. Longman, London. A very clear, detailed and well illustrated discussion of the process of translation.

COLD SPRING HARBOR SYMPOSIA OF QUANTITATIVE BIOLOGY: these appear annually, and many comprise collections of topical papers on replication, transcription and translation.

LOSICK, R. (1972) *In vitro* transcription. *Annual Reviews of Biochemistry*, **41**, 409–446. Illustrates the progress in understanding transcription and its regulation.

Chapter 7

JACOB, F. & MONOD, J. (1961) Genetic regulatory mechanisms in the synthesis of proteins. *Journal of Molecular Biology*, **3**, 318–356. Classical paper on the regulation of enzyme synthesis.

REZNIKOFF, W. S. (1972) The operon revisited. *Annual Reviews of Genetics*, **6**, 133–156. More extensive discussion of operon systems in bacteria.

BECKWITH, J. R. & ZIPSER, D. (1970) *The Lactose Operon*. Cold Spring Harbor Laboratory. A book completely devoted to the *lac* operon containing ten very useful introductory chapters and other original papers.

MANDELSTAM, J. & McQUILLEN, K. (1973) *Biochemistry of Bacterial Growth*. Blackwell Scientific Publications, Oxford. Chapter 8 on coordination of metabolism very useful.

Chapter 8

ASHWORTH, J. M. & SMITH J. E. (editors) (1973) Microbial Differentiation. *Symposium of the Society for General Microbiology*, **23**. Cambridge University Press, Cambridge. A collection of review articles covering the main morphogenetic systems presently under study.

Critical Reviews in Microbiology, **1**, 3. (1972). An issue devoted to research articles on different microbial morphogenesis systems.

MANDELSTAM, J. (1969) Regulation of bacterial spore formation. *Symposium of the Society for General Microbiology,* **19,** 377—402. Discusses the biochemistry of spore formation in bacteria.

HERSHEY, A. D. (editor) (1971) *The Bacteriophage Lambda.* Cold Spring Harbor Laboratory. Begins with fifteen introductory chapters followed by original research papers, all devoted to this phage.

KUSHNER, D. J. (1969) Self-assembly of biological structures. *Bacteriological Reviews,* **33,** 302—345. Covers assembly of phage, other viruses, bacterial flagella, pili, microtubules, ribosomes, membranes and wall structures.

WOOD, W. B. & EDGAR, R. S. (1967) Building a bacterial virus. *Scientific American,* July 1967. Well illustrated account of the mechanism of assembly of phage T4.

Abbreviations

Amino acids

ala	alanine	glu	glutamic acid
DAP	diaminopimelic acid	lys	lysine
DPA	dipicolinic acid	ser	serine

Cofactors

ACL	antigen carrier lipid	NAD	nicotine adenine dinucleotide
ACL.P	antigen carrier lipid phosphate	NADP	nicotine adenine dinucleotide phosphate
ACP	acyl carrier protein		
CoA	coenzyme A	PAB	*para*-aminobenzoic acid
FMN	flavin mononucleotide	TPP	thiamine pyrophosphate
fd	ferredoxin		

Nucleosides and nucleotides

AMP	adenosine monophosphate	APS	adenylyl sulphate
ADP	adenosine diphosphate	PAPS	phosphoryladenylyl sulphate
ATP	adenosine triphosphate	PRPP	phosphoribosyl pyrophosphate
CMP	cytosine monophosphate	TTP	thymine triphosphate
CDP	cytosine diphosphate	UMP	uridine monophosphate
CTP	cytosine triphosphate	UDP	uridine diphosphate
GMP	guanosine monophosphate	UTP	uridine triphosphate
IMP	inosine monophosphate		

Sugars (Monosaccharides)

gal	galactose	KDO	ketodeoxyoctonic acid
glc	glucose	man	mannose
glcNac	*N*-acetylglucosamine	murNAc	*N*-acetylmuramic acid
hept	heptose	rha	rhamnose

Others

DNA	deoxyribonucleic acid	P_i	orthophosphate
RNA	ribonucleic acid	PP_i	pyrophosphate
*m*RNA	messenger RNA	PHB	poly-β-hydroxybutyric acid
*t*RNA	transfer RNA	TCA	tricarboxylic acid (Krebs) cycle
*r*RNA	ribosomal RNA		

Glossary

allosteric protein   A protein, usually an enzyme, which undergoes an alteration in conformation on binding low molecular weight molecules (ligands). The binding alters the (catalytic) activity of the protein (see p. 150).

autotroph   An organism which does not require any organic carbon for its energy source or for growth.

auxotroph   An organism requiring a particular organic molecule for growth. Often used to refer to mutants with a specific requirement such as an amino acid.

cistron   The functional unit of hereditary, which in physical terms is a region of DNA coding for a single polypeptide chain or ribosomal or transfer RNA species. The concept is similar to that of a *gene*, but is used where a more specific term is required.

heterokaryosis   The process whereby cells fuse to form a multi-nucleate cell containing nuclei from different parents. Nuclear fusion does not take place.

heterozygosis   The process whereby cells and their nuclei fuse to form a cell containing a single nucleus with chromosomes from each parent.

lysogen   A bacterium containing a bacteriophage chromosome integrated into and replicating with its own chromosome.

meiosis   The processes of nuclear division associated with the formation of gametes or of haploid cells from a diploid.

mitosis   The normal process of nuclear division in a eukaryote whereby nuclear division occurs on a spindle structure without a reduction in the chromosome number in the daughter nuclei.

Park nucleotides   Bacterial cell wall precursors (including UDP-*N*-acetylmuramic acid, and its amino-acyl derivatives) which are found to accumulate in cultures in the presence of certain inhibitors of cell wall synthesis.

poly(A)   A polymer of adenylic acid residues found attached terminally to messenger RNA fractions.

plasmid   A piece of extrachromosomal DNA capable of replication independent of the host chromosome.

spindle plaque   A structure in some lower eukaryotes replacing the centriole. The spindle plaque lies on the nuclear membrane and from it radiates the spindle fibres on which chromosomes segregate during meiosis or mitosis.

temperate bacteriophage   A bacteriophage with the alternative fates after infection of either integrating its DNA into the host chromosome to form a *lysogen*, or of undergoing lytic development.

Index